PROCESS EQUIPMENT SERIES
VOLUME 4

Solids and Liquids Conveying Systems

Edited by
Mahesh V. Bhatia, P.E.
Uniroyal Chemicals
Naugatuck, CT

Contributors to Volume 4

J. R. Birk
Donald T. Deeley
J. L. Harris
K. S. Panesar
Rocco J. Siclare
R. E. Syska
W. W. Willoughby

PROCESS EQUIPMENT SERIES
VOLUME 4

Solids and Liquids
Conveying Systems

©Technomic Publishing Company, Inc. 1982
P.O. Box 913, 265 Post Road West, Westport, CT 06881

Printed in U.S.A.
LCC 79-63114
ISBN 87762-310-4

FOREWORD

This is the fourth volume in the *Process Equipment Series* which is a presentation of useful information relating to equipment and devices used in the chemical and related process industries. The editorial material has been prepared and written by specialists in their respective fields. The presentation has been divided among a series of volumes.

The equipment described in this series is used by more than 10,000 chemical manufacturing and processing firms, employing 1.1 million Americans, with sales of approximately 100 billion dollars. Chemical industry products represent 7.5 percent of the U.S. Gross National Product (GNP) and enjoy an annual growth rate nearly twice that of the overall GNP.

Equipment, apparatus and devices employed by the process industries are so numerous and varied that the problem was to reduce such information to a moderate and practical size. The editor has given preference to equipment that has general application, and diverse uses in the process industries. Some of the vacuum devices useful to transfer liquids are discussed in the third volume of the series (Air Movement and Vacuum Devices).

Material presented by authors is primarily written in general technical language and the text liberally supplemented with drawings and photographs. The intent has been to provide sufficient theoretical matter to the reader with a satisfactory understanding of the equipment or devices covered. Equipment discussed is of the type that can be purchased by contrast to in-house-designed or constructed apparatus.

The editor expresses his sincere thanks to the experts who contributed to these collected volumes.

Mahesh V. Bhatia

TABLE OF CONTENTS

CHAPTER 1

SOLIDS CONVEYING

ROCCO J. SICLARE, P.E.
PME Equipment, Inc.
Union, New Jersey

INTRODUCTION

The early processing and milling operations were designed so that material flowed through the plant by gravity. Plants had to be self-sufficient and were located close to all the raw materials required for the operation. The bulk granular raw materials were carried to the plant by wagon and unloaded at a high point on the terrain for gravity feed into the plant. Where materials could not be moved by gravity, they were manually shovelled, pushed, carried, or lifted into the processing unit. The operations were batch-type and fortunately required only an intermittent flow of materials.

During the last seventy-five years many conveyors, such as the belt conveyor, screw conveyor, vibratory conveyor, and pneumatic conveyor, have been developed for the handling of lumpy, granular or fine materials and this technology has played a major role in the development of modern industry and the advancement of man. These conveyors move materials into the plant, through the various unit operations and out of the plant; swiftly, smoothly, safely and economically. They convey all types of bulk materials over short or long distances, in and out of storage, and through the many unit operations of a processing plant. They have made possible the transition from small process batch plants to the large continuous processes of today's modern technology.

A material handling system can be a single conveyor or it can be comprised of many conveyors of the same or different types to meet the conveying and unit operational requirements of the process. The selection of the most economical conveying equipment is normally dependent, to a large degree, on the characteristics of the material to be conveyed, the capacity of the conveyor, the physical conveying requirements of the conveyor and, to a lesser degree, on the process requirements of the conveyor.

In order to properly select and size a conveyor, the characteristics of the material to be conveyed must be known and studied. Characteristics such as particle size, flowability and abrasiveness of the material can often be the determining factors in selecting and sizing a conveyor. Other characteristics, such as the corrosiveness of the material, may require careful consideration and be the dominating factor for, perhaps, the selection of a belt conveyor as a suitable and economically practical conveyor for this application. Some materials may have fire or explosive hazards which would certainly require

1

special attention. This may dictate the use of a pneumatic conveyor with an inert gas as the carrier. Some materials may be poisonous, dusty, or release fumes which may be undesirable or hazardous to the operators, and this operation may require the use of a totally enclosed type conveyor, such as a screw conveyor. There are also some materials, such as foods, that are subject to spoilage, and this may require conveyors that can be easily cleaned and that are of sanitary design, such as vibratory conveyors.

MATERIALS CLASSIFICATIONS

Some years ago, a system of material classification was developed by the Link Belt Division of FMC which allows for the convenient classification of any bulk material according to its characteristics. This method is shown in Table 1.1. For example, a material which is fine, free-flowing, non-abrasive and packs under pressure, would fall under classifications B, 2, 6 and Z and have a B26Z classification.

Table 1.1. Material Class Description

	Material Characteristic	Class
Size	Very fine—100 mesh and under	A
	Fine—1/8-inch mesh and under	B
	Granular—½-inch and under	C
	Lumpy—containing lumps over ½ inch	D
	Irregular—being fibrous, stringy, or the like	H
Flowability	Very free flowing—angle of repose up to 30°	1
	Free flowing—angle of repose 30° to 45°	2
	Sluggish—angle of repose 45° and up	3
Abrasiveness	Nonabrasive	6
	Mildly abrasive	7
	Very abrasive	8
Other characteristics	Contaminable, affecting use or saleability	K
	Hygroscopic	L
	Highly corrosive	N
	Mildly corrosive	P
	Gives off dust or fumes harmful to life	R
	Contains explosive dust	S
	Degradable, affecting use or saleability	T
	Very light and fluffy	W
	Interlocks or mats to resist digging	X
	Aerates and becomes fluid	Y
	Packs under pressure	Z

Table 1.2. Classification of Bulk Materials

Material	Class	Bulk Density lbs./Cu.Ft.
Alum, Lumpy	D26	50-60
Alum, Pulverized	B26	45-50
Alumina	B28	60
Aluminum Hydrate	C26	18
Aluminum Oxide	D28	75-85
Asbestos, Shred	H37WZ	20-25
Bagasse	H36X	7-8
Bakelite, Powder	A36	30-40
Bauxite, Crushed	D28	75-85
Bonemeal	B27	55-60
Borox, Powder	B26	53
Cement, Portland	A27Y	75-85
Cement, Clinker	D28	40
Coal, Anthracite, Granular	B28P	60
Coal, Bituminous, Minus ½″	D26P	50
Cocoa, Powder	A26Z	30-35
Cocoa, Beans	C27T	30-40
Coffee, Green Bean	C26T	32
Coffee, Ground	B26	25
Coffee, Roasted Bean	C16	22-26
Coke, Loose	D28TX	23-32
Copper, Ore	D28	120-150
Cork, Fine Ground	B36WY	12-15
Corn, Cracked	C16	45-50
Cullet	D28	80-120
Dolomite, Crushed	D27	90-100
Fly Ash	A14Y	35-40
Graphite, Flour	A16Y	28
Gypsum, Calcined	C27	53-60
Ice, Crushed	D16	35-45
Lime, Hydrated	B26YZ	40
Lime, Pebble	D26	56
Limestone, Crushed	D27	85-90
Oats	C16S	26
Peanuts, Shelled	C26	20-25
Phosphate Rock	D27	75-85
Rice, Hulled	B16	45
Salt, Fine	B26PL	70-80
Sand, Dry	B18	90-100
Soap Powder	B26	20-25
Soda Ash, Light	A27W	20-35
Soda Ash, Heavy	C27	55-65
Sodium Sulfate	B26	45
Stone, Crushed	D28X	85-90
Sugar, Granulated	B26KT	50-55
Wheat	C16S	45-48

Table 1.2 lists many of the materials handled by conveyors and includes the material classification along with the average bulk density. This system of classification was recently expanded by the Conveyor Equipment Manufacturers Assocation, (CEMA), and listed in their Book No. 550. Table 1.3 is a guide for the preliminary selection of a conveyor based on material characteristics.

CAPACITY & SELECTION

The capacity of conveyors is normally expressed in pounds or tons per hour. Since mechanical conveyors are volumetric machines, for sizing purposes, capacities are converted to cu. ft. per hour using the bulk density of the material. In addition to the average hourly capacity requirements of the conveyor, sizing considerations must include any peak loads that may be caused by the inherent design or natural surging or batching ofother equipment in the system. The peak loads are considered to be the instantaneous rate requirements of the conveyor and are normally the basis for sizing. In addition to the hourly and peak rates, a conveyor design should also be able to handle rates and loads which may occur due to an emergency upset condition in the process.

In most applications the flow of material into a conveyor is controlled by the use of a volumetric feeder. Feeders are available based on most conveyor types and include belt, screw, vibratory, apron, rotary vane, plow and table type feeders. Gravimetric weigh belt feeders are available where feeding accuracy is required and a weight measurement is desirable.

Conveyors are available for conveying materials horizontally, vertically or on a slope. Belt conveyors can be designed for the transport of materials over considerable distances, both horizontal and sloped as demonstrated by the 60 inch wide by 30 mile long multi-belt conveyor system operating in Morocco, handling phosphate rock and the 42 inch wide by 10 mile long belt system operating in Ohio handling coal. Screw conveyors are usually limited to maximum horizontal lengths of 150 feet, and vibratory conveyors up to lengths of 300 feet. Pneumatic conveyors usually do not exceed 700 feet. Bucket elevators can be designed for vertical lifts up to 200 feet. Belt conveyors are limited to lifts and drops determined by the angle of travel and the available space. The space requirements of a conveyor can often be the determining factor in the selection of a suitable conveyor.

In addition to the transporting function, some conveyors can simultaneously perform unit operations which are required by the system. Screw conveyors can be jacketed for the cooling and heating of materials. They can also be used for dewatering operations and fitted with special flighting and paddles for mixing materials. Vibratory conveyors can also be used to heat and cool mate-

Table 1.3. Conveyor Selection Guide

Type of Conveyor	Size Fine	Size Granular	Size Large Lumps	Flowability Very Free	Flowability Free	Flowability Sluggish	Abrasiveness Non-AB	Abrasiveness Mildly AB	Abrasiveness Very AB	Path Horizontal (H)	Path Inclined (I)	Path Vertical (V)	Path Combined (HI)	Path Combined (HV)
Belt conveyor	x	x	x	x	x	x	x	x	x	x	x		x	
Screw conveyor	x	x		x	x	x	x	x	x	x	x			
Vibratory conveyor		x	x	x	x		x	x	x	x				
Bucket elevator	x	x			x		x	x			x	x		
Apron conveyor		x	x	x	x	x	x	x	x	x	x		x	
Drag conveyors	x	x	x	x	x	x	x	x		x	x		x	
Pivot bucket elevator		x	x	x	x		x	x	x	x			x	x
Pneumatic conveyor	x	x		x	x		x	x		x		x		x

rials and are easily adapted with screens and grizzleys for particle sizing and scalping operations. The ability to perform these process functions may at times determine the selection of a conveyor for a particular conveying requirement.

There are approximately twenty-four conveyors used in the process industries. This section discusses only the most frequently used conveyors, belt, screw, vibratory, bucket and pneumatic, and the section is geared primarily to the use and selection of these conveyors for the process industries.

SCREW CONVEYOR

The screw conveyor is one of the most widely used conveyors in the process industries. Flow of material is accomplished by the use of a rotating helix inside of a fixed trough. The rotating helix or flighting provides a continuously inclined face which creates a forward movement of the product.

There are very few fine or granular products which can not be conveyed by a screw conveyor. The stationary and enclosed dust tight design of the trough housing make the conveyor ideally suited for conveying both large and small quantities of dusty products over long or short distances. Conveying distances are usually held to within 150 feet and restricted by the torque limitations of the conveyor components.

Screw conveyors can be fabricated in essentially all materials of construction, such as stainless steel, aluminum, titanium and monel, in addition to carbon steel and special abrasion resistant forms of carbon steel. They can also be rubber lined and hot dipped galvanized to meet the special abrasion and corrosion requirements of modern industry.

The stationary design of the trough makes it an excellent conveyor for applications where multiple feed points and discharge points are required. The feed can be directed to any one of the outlets, whether it be upstream or downstream of the feed point, by reversing the rotation of the screw. In addition, by use of the hand of the screw, right hand or left hand, as shown in Figure 1.1, it is also possible to use the conveyor for transporting material in opposite directions at the same time. These features are unique with a screw conveyor and give the design engineer layout options which are not normally available with other mechanical conveyors.

Screw Conveyor Components

The heart of the screw conveyor is the flighting, and this is available in two forms, as shown in Figure 1.2. The helicoid type is a continuous flight and is formed by a cold rolling operation from strip steel. The resulting flight is one piece construction having a tapered cross section so that the outer edge of the flight is one-half the thickness of the inner edge. In addition, the rolling process produces a smooth hardened surface which has greater wear resistance than the original material.

6

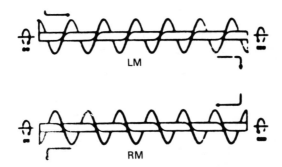

Figure 1.1 Conveyor screw rotation.

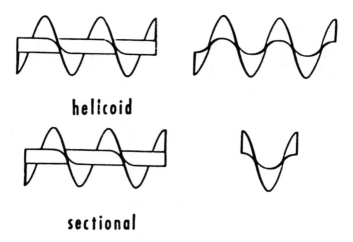

helicoid

sectional

Figure 1.2. Types of flighting.

The second type of flighting is sectional flighting and these are press formed from flat plate blanks. Each blank, which is in the shape of a donut, produces one flight and the flights are then manually butt welded to each other to form a continuous ribbon flight. Sectional flights have a uniform thickness. In both the helicoid and sectional types, standard flighting is formed so that the pitch is equal to the diameter of the flight.

Screw flighting is availabe in many variations, as shown in Figure 1.3 for special applications. Ribbon flights are available for handling wet and sticky products such as filter cake. Cut flights, cut and folded flights and paddles are used in mixing applications. Tapered flights and variable pitch flights are available for feeding applications and where uniform product withdrawal from a hopper or bin is desirable.

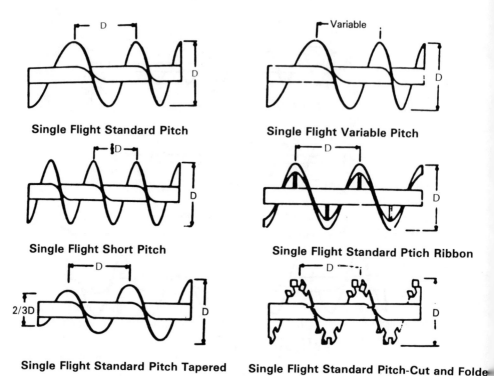

Single Flight Standard Pitch

Single Flight Variable Pitch

Single Flight Short Pitch

Single Flight Standard Ptich Ribbon

Single Flight Standard Pitch Tapered

Single Flight Standard Pitch-Cut and Folde

Figure 1.3. Types of screws.

Helicoid and sectional flightings are available in standard sections of 9′ −10″ for screws up to 9 inch diameter and in 11′ −9″ sections for larger screws. Coupling shafts are used to join these sections to form longer screws. The couplings are supported in bearings which are mounted in hanger assemblies. Common bearing materials are hard iron, babbit, wood, bronze, stellite, reinforced plastics, and manganese steel. These bearings are usually operated without any lubrication. Couplings are available in standard cold roll steel and surfaced hardened steel, in addition to stainless steel and other alloys.

Screw troughs are usually "U" or flared shaped and designed for a ½″ clearance between the screw and trough as shown in Figure 1.4. The "U" trough is used in most applications. The flange of the trough is usually formed using a structural angle. The flange of the trough can also be formed from the same sheet as the trough and this design is sometimes used in stainless steel construction, and for air-tight and sanitary construction. The flared troughs are normally used where the screw forms the bottom of a hopper, such as in feeder applications, and for handling some sticky products.

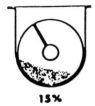 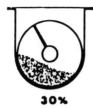  

15%   30%   45%   100%

Figure 1.4. Conveyor trough loadings.

For dust tight construction the trough cover is a critical item. Covers are usually flanged and a minimum of 12 ga. in thickness. Gasket material is normally soft, of less than 40 durometer, and a minimum of ¼ " in thickness so that it conforms to the uneven surface of the cover and trough flange.

Size Determination

Screw conveyor sizing is a function of the conveying rate, screw speed and screw capacity based on the cross-sectional trough loading. Table 1.4 gives the capacity of screws up to 24″ diameter at 15, 30, 45 and 100 percent trough loadings, as shown in Figure 1.4. The capacities, expressed in cubic feet per hour per rpm of screw speed, are theoretical capacities and calculated from the formula:

$$C_s = \frac{0.785\,(D_s{}^2 - D_p{}^2)\,PK\,60}{1728}$$

(Equation 1.1)

where:

C_s = Screw Capacity, cu. ft. per hour per rpm.
rpm = revolutions of screw per minute
D_s = Diameter of screw, inches
D_p = Outside Diameter of pipe, inches
P = Pitch of screw, inches
K = percent of trough loading, fraction

The formula is based on the assumption that for each revolution of the screw, the material contained within one pitch will travel one pitch distance. In addition, it also assumes that there is no movement of material in the half inch clearance between the trough and screw. The listed capacities are acceptable basic design criteria for screw conveyors and are used for essentially all applications.

For moderately abrasive materials, and most materials do fall into this

9

Table 1.4. Capacity Chart & Recommended Operating Speeds

| Screw | Capacity-Cubic Feet Per Hour Per RPM | | | |
| Diameter-Inches | Trough Loading-Percent | | | |
	15	30	45	100
4	0.21	0.4	0.6	1.3
6	0.75	1.5	2.3	4.8
9	2.8	5.6	8.0	17
10	3.7	7.2	10.9	24
12	6.7	13.3	19.5	44
14	10.8	21.1	31.0	68
16	15.9	31.5	46.6	104
18	22.7	45.7	66.3	150
20	31.2	62.4	95.6	208
24	58.0	118.0	172.0	340
Recommended Operating Speeds	20–35	35–55	55–90	20–100

general classification, a trough loading of 30 percent is normally used in the design of screw conveyors where an internal bearing is required. Operation at this level allows the screw to convey the material below the bearing, thereby reducing wear and frequent maintenance of the bearing. For some materials, primarily agricultural and food, such as roasted coffee beans, the screw can be selected based on 45 percent trough loading. For very abrasive materials, such as alumina, cullet and sand, trough loadings of 15 percent are not uncommon. In applications where an internal bearing is not required and bearing maintenance is no longer a consideration, the loadings can be increased to 60–70 percent. Trough loadings up to 100 percent can also be cautiously considered for this case.

Table 1.4 also lists recommended screw speeds for all sizes based on cross sectional loading. These speeds are significantly lower than those found in the literature. However, they do represent acceptable speeds for conveyors operating continuously, 24 hours per day. The recommended speeds for 15, 30 and 45 percent loading assumes that the screw has at least one internal hanger bearing. For screws without internal bearings, screw speeds can be increased to the levels shown for 100 percent loading. In general, when handling moderately abrasive products, operating speeds for screw conveyors with internal bearings should fall between 35 and 55 rpm. For handling very abrasive products screw speeds should fall between 20 and 35 rpm, and sometimes where maintenance and conveyor failure is of major concern, down to 15 rpm. On applications where there are size limitations due to equipment clearances, openings into reactors, etc., then of course, these speeds can be increased.

The screw speed is calculated using the formula:

$$S_c = \frac{C_r \times C_f}{C_s}$$

(Equation 1.2)

where:

S_c = Screw Speed, rpm
C_r = Conveying Rate, cu. ft. per hour
C_s = Capacity of screw, cu. ft. per hour per rpm, from Table 1.4
C_f = Speed Correction Factor, from Table 1.5

Table 1.5. Speed Correction Factor

Type of Flight	C_f
Standard Pitch	1.0
Short Pitch	1.50
Half Pitch	2.0
*Standard Ribbon	1.37
*Cut Flight	1.57

*At 30% loading.

For standard screws, where the pitch is equal to the diameter of the screw, C_f is unity. Correction factors are given in Table 1.5 for flighting configurations other than standard. As used in the above formula, these factors adjust the speed of the screw, correcting for the inefficiencies of the cut and ribbon flighting and the reduced screw capacities for pitches less than standard. For any flight pitch the correction factor can be calculated using:

$$C_f = \frac{D_s}{P}$$

(Equation 1.3)

Sample Calculation: Select a screw conveyor for conveying 20,000 pounds per hour of hydrated lime, weighing 40 pounds per cubic foot, from the discharge of a rotary feeder a horizontal distance of 30 feet to the inlet of a blender.

Solution: Hydrated lime is considered to be moderately abrasive and since the length of the conveyor will require an internal hanger bearing, a screw loading of 30 percent will be used for this calculation. In addition, for 30 percent loading, a screw should be selected that will operate between 35—55 rpm. Therefore, conveying rate will be:

$$C_r = \frac{20,000}{40} = 500 \text{ CFH}$$

Assuming a 9″ screw which has a capacity of 5.6 cubic feet per hour per rpm, the screw speed will be:

11

$$S_c = \frac{500 \times 1}{5.6} = 89 \text{ rpm}$$

Since 89 rpm falls beyond the 35—55 speed range, assume a larger size or 12" dia. A 12" dia. screw at 30 percent loading has a capacity of 13.3 cubic feet per hour per rpm. The modified screw speed will be:

$$S_c = \frac{500 \times 1}{13.3} = 38 \text{ rpm}$$

Therefore, a 12" dia. screw is required to feed 20,000 pounds per hour of hydrated lime.

Screw conveyors are often used on an incline. Flight capacities are reduced since there is a reduction in the effective angle of the flight and the resulting positive forces required to push the material forward. Figure 1.5 gives the efficiencies of screw conveyors as a function of angle of incline. In general, as the screw is inclined above 7 degrees, a decline in efficiency of 2 percent can be expected for every degree of incline of the screw up to 45 degrees. This will vary depending on the characteristics of the material conveyed with the fine and fluid products being the most troublesome. The formula for screw speed then becomes:

$$S_c = \frac{C_r C_f}{C_s E} \qquad \text{(Equation 1.2a)}$$

where E is the efficiency factor from Figure 1.5 for standard pitch screws. Normally screw conveyors should be limited to inclines of 15 degrees since a screw can handle most materials at this angle. Short pitch screws are recommended for difficult materials and designs without hanger bearings are preferable. Cut flights and ribbon flights are seldom used.

In the previous sample calculation if the screw was inclined 15 degrees the screw efficiency from Figure 1.5 is 71 percent. The calculated speed would then be 37 rpm/0.71 = 52 rpm minimum.

Horsepower Determination

The horsepower required to drive a screw conveyor is arrived at by calculating the power required to move the material horizontally and the power required to overcome the frictional resistance in the internal components of the conveyor.

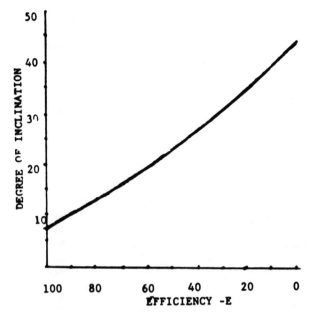

Figure 1.5. Effect of incline on screw conveyor capacity.

$$\text{a)} \qquad HP_m = \frac{C_w \, L \, F_m}{1 \times 10^6} \qquad \text{(Equation 1.4)}$$

where:

HP_m = Material Horsepower
C_w = Flow Rate, pounds per hour
L = Conveyor Length, feet
F_m = Material Horsepower Factor from Table 1.6

$$\text{b)} \qquad HP_f = \frac{L \, S_c \, F_c \, F_b}{1 \times 10^6} \qquad \text{(Equation 1.5)}$$

where:

HP_f = Friction Horsepower
S_c = Conveyor Speed, rpm
F_c = Diameter Factor, from Table 1.7
F_b = Bearing Factor, from Table 1.8

Table 1.6. Bulk Material Design Data

Material	Bulk Density Lbs/Cu. Ft.	Recommended Conveyor Loading	F_m Material H.P. Factor
Alum, Lumpy	50-60	30	1.4
Alumina	60	15	1.8
Ammonium Chloride	52	30	0.7
Asbestos, Shred	20-25	30	1.0
Ashes, Dry	35-40	30	4.0
Asphalt, Crushed	45	30	2.0
Bakelite, Powdered	30-40	30	1.4
Bark, Wood	10-20	30	1.2
Bauxite, Crushed	75-85	15	1.8
Bones, Crushed	35-40	30	2.0
Bonemeal	55-60	30	1.7
Borax, Powdered	53	30	0.7
Carbon Black, Pellets	25	30	0.4
Carbon Black, Powder	4-6	30	0.4
Cement	75-85	30	1.4
Chalk	85-90	30	1.9
Charcoal	18-25	30	1.4
Coal	50	30	1.0
Cocoa	30-35	30	0.9
Cocoa Beans	30-40	30	0.4
Coffee Beans	32	30	0.8
Coffee, Ground	25	30	0.4
Cork, Fine	12-15	30	0.5
Corn, Seed	45	45	0.4
Corn, Sugar	31	30	1.0
Cottonseed	35	30	0.9
Cullet	80-120	15	2.0
Dolomite	90-100	30	2.0
Feldspar	65-70	30	2.0
Flour, Wheat	35-40	45	0.6
Flue, Dust	110-125	30	3.0
Fly Ash, Dry	35-40	30	3.0
Foundry Sand	90-100	30	2.0
Fuller's Earth	40	15	2.0
Gypsum	90-100	30	1.6
Ice, Crushed	35-40	30	0.4
Lead, Oxide	30-150	30	1.0
Lime, Hydrated	40	30	0.8
Lime, Pebble	56	30	1.3
Limestone, Crushed	85-90	30	2.0
Mica, Ground	13-15	30	0.7
Paper Pulp 6-15%	60-62	30	1.2
Phosphate, Rock	75-80	30	1.8
Rice, Hulled	45-48	45	0.4
Rice, Grits	42-45	30	0.4

(Continued)

Table 1.6. Bulk Material Design Data (Continued)

Material	Bulk Density Lbs/Cu.Ft.	Recommended Conveyor Loading	F_m Material H.P. Factor
Rubber, Ground	23	30	0.8
Salt, Dry Fine	70-80	30	1.0
Salt, Cake	90-100	15	2.0
Sand, Dry, Silica	90-100	15	2.0
Sawdust	10-13	30	0.7
Shavings, Wood	15	30	1.0
Soap, Powder	20-25	30	0.9
Soda Ash, Heavy	55-65	30	0.7
Soybeans, Whole	45-50	30	0.4
Sugar, Granulated	50-55	30	0.7
Sulfur, Crushed	50-60	30	0.6
Tri-Sodium Phosphate	60	30	1.7
Wheat	40-48	45	0.4

(Concluded)

Table 1.7. Conveyor Diameter Factor

Diameter	Factor F_c
4	12
6	18
9	31
10	38
12	55
14	78
16	107
18	139
20	165
24	226

c)
$$HP_r = \frac{(HP_m + HP_f) F_o}{e}$$
(Equation 1.6)

where:

HP_T = Conveyor Total Horsepower
F_o = Overload Factor, from Figure 1.6
e = Drive Efficiency, from Table 1.9

The conveyor overload factor, F_o, corrects the formula for temporary over-

Table 1.8. Hanger Bearing Factor

Bearing Type	Factor F_b
Arguto	1.7
Babbit	1.6
Ball	1.0
Bronze	1.7
Lignum Vitae	1.7
Manganese Steel	4.4
Nylatron	1.7
Oilite	1.7
Stellite	4.3
Teflon	2.0
Hard Iron	4.4
Wood	1.8
Gatke	2.0

Table 1.9. Drive Efficiencies

Type	Efficiency Factor
V-Belt and Sheaves	0.94
Roller Chain and Sprocket	0.93
Shaft Mounted Gear Reducers, Helical Gear	0.94
Foot Mounted Gear Reducers, Helical Gear	0.94
Shaft Mounted Gear Reducers, Worm Gear	0.70
Foot Mounted Gear Reducers, Worm Gear	0.70
Cut Tooth Spur Gears, Open for Each Reduction	0.90

loads that may occur during start-up of the screw conveyor. For horsepowers greater than 5 HP, F_o is equal to 1.0.

It should be noted that the horsepower formulas are empirical and are only applicable for calculating the power requirements of screws handling dry materials. The power required to move wet or moist materials may be appreciably higher due to the added friction resulting from a build-up of hard non-flowing material in the trough. Formulas are not available for approximating the power requirements for wet materials.

The horsepower required for an inclined screw can be calculated by adding the lift horsepower to the material and friction horsepower as indicated above, and making a correction for the inefficiencies of the inclined screw, as follows:

$$HP_T = \frac{(HP_m + HP_f + HP_h) F_o}{e \times E} \qquad \text{(Equation 1.6a)}$$

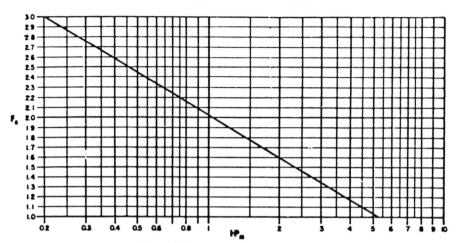

Figure 1.6. Horsepower overload factor.

where:

$$HP_h = \frac{Cw\ H}{1 \times 10^6}$$ (Equation 1.7)

and where:

HP_h = Conveyor Lift Horsepower
Cw = Flow Rate, pounds per hour
H = Actual Height of Lift, feet
E = Efficiency Factor, from Figure 1.5

The torque developed by the drive to rotate the screw is transmitted through the screw conveyor components. The drive torque is calculated using the formula:

$$T_D = \frac{HP_D \times 63,000}{S_c}$$ (Equation 1.8)

where:

T_D = Drive Torque, pounds
HP_D = Drive Horsepower
S_c = Screw Speed, rpm

To prevent failure of the screw conveyor components, the various parts are sized and selected to transmit at least the maximum torsional requirements of the screw as calculated above. Table 1.10 lists the torsional ratings of standard carbon steel conveyor components and the ratings are listed to correspond

Table 1.10. Torsional Rating of Couplings, Shafts, Pipes and Bolts

Coupling and Shafts			Pipe			Bolts
	Torque, lbs.					Torque, Shear, lbs. 2-Bolt
Dia	Std.	Hard	Dia	Torque, Lbs	Dia	Design
1	820	1025	1 ¼	3140	3/8	1380
1 ½	3070	3850	2	7500	1/2	3660
2	7600	9500	2 ½	14250	5/8	7600
2 7/16	15050	18900	3	23100	5/8	9270
3	28370	35400	3 ½	32100	3/4	16400
3	28370	35400	4	43000	3/4	16400
3 7/16	42550	53000	4	43000	7/8	25600

with standard sizes based on shaft size. The screw components are selected from components having a greater torsional rating than the drive.

BELT CONVEYOR

Belt conveyors have a wide range of capacities, are extremely dependable and normally require a minimum amount of maintenance. They are the workhorses of the conveyor industry and, as a rule of thumb, receive first consideration in the analysis of any material handling problem. In addition to low maintenance, their power requirements are significantly lower than other conveyors.

They are used mostly for handling lumpy, granular or irregularly shaped materials over short to very long distances. They are seldom used to handle fine, dusty materials except where conveying distances greater than 1000 feet are involved, or where capacity requirements are high and beyond the range of pneumatic, screw, or drag type conveyors. On occasion they are used to handle fine wet materials, such as filter cake, but this application requires a higher degree of maintenance than is usually necessary. Since the material is stationary once loaded on the belt, belt conveyors are excellent conveyors for handling abrasive or fragile materials and materials which cannot tolerate any metallic contamination.

Belt Conveyor Components

Conveyor belting is available in 14, 16, 18, 20, 24, 30, 36, 42, 48, 54, 60, 66, and 72 inch wide sizes. The belt essentially consists of rubber or PVC top and bottom covers in various thicknesses with the carrying side cover thicker than the underside cover. The interior of the belt is the carcass which consists of a number of layers of cotton, rayon, nylon or polyester fabric with rubber support layers in between the plies. The carcass of the belt is solely responsible for the strength of the belt.

For short to intermediate belt conveyors the effective belt tension may be calculated using the formula

$$T_E = \frac{HP_T \times 33,000}{S} \qquad \text{(Equation 1.9)}$$

where T_E = Effective Belt Tension, pounds.

The maximum tension in the belt is greater than the effective belt tension due to the additional tension created by the take up load. The maximum belt tension for single pulley drives may be calculated by multiplying T_E by a factor of 1.5 for gravity type take-ups and 2.0 for screw type take-ups. Belts are normally rated in pounds per inch of width and are selected based on the maximum tension.

The loaded belt is supported on carrying idlers which in most applications are equally spaced along the length of the conveyor. The distance between idlers is a function of belt width and product density. Table 1.11 gives recommended idler spacings based on an average belt sag between idlers of 2 to 3 percent of the distance between idlers. The sag will vary based on belt tension and consequently is greatest at the feed end and least at the discharge end. In practice, with medium length belts, it is much simpler to use an average spacing rather than vary the spacing to accommodate sag. However, usually idlers are spaced closer at the feed end to absorb the impact forces and to maintain the contour of the belt.

The empty belt is supported on return idlers which are normally spaced at 10 feet. Return idlers can be a simple steel roll type or a rubber disc type for handling sticky materials. Build-up of material on the return idlers can cause the belt to shift and contact the idler bracket. This type operation can destroy the belt.

Belt take-ups are either manual or automatic gravity type take-ups. Both are used to apply the necessary minimum slack side tension required for proper operation of the belt and for releasing the belt tension for maintenance purposes. The screw type manual take-up is normally used for belt lengths less than 100 feet and is located at the tail pulley.

Automatic gravity take-ups are either the vertical or horizontal type. In both cases a take-up weight is used to apply the necessary minimum belt tension and this tension is constant. An increase in belt length due to tension is automatically taken up. The vertical type is usually located close to the drive end of the conveyor. The horizontal type is located at the tail end and utilizes

Table 1.11. Troughing Idler Spacing

Belt Width	Recommended Idler Spacing Inches				
	Material Weight, Lbs. Per Cu. Ft.				
Inches	25	50	75	100	150
14	66	60	60	60	54
16	66	60	60	60	54
18	66	60	60	60	54
20	66	60	54	54	48
24	60	54	54	48	48
30	60	54	54	48	48
36	60	54	48	48	42
42	54	54	48	42	36
48	54	48	48	42	36
54	54	48	42	42	36
60	48	48	42	36	36
66	48	48	42	36	36
72	48	42	42	36	30

the tail pulley in combination with a floating weight and cable to apply the necessary tension.

A belt tripper is used when it is desirable to discharge material at points over the length of the conveyor. The design essentially consists of a movable raised head pulley which lifts the belt so that the product is discharged into a chute mounted on the tripper. Trippers can either be moved by hand or are power driven. Power driven units operate at speeds up to 60 feet per minute. When travelling in a direction opposite to belt travel, the tripper speed must be added to the belt speed to arrive at the correct instantaneous discharge rate. Depending on the belt speed, this can be a sizable increase in discharge rate.

In recent years cleaning material from the dirty side of the belt as it leaves the head pulley has taken on an added importance. Failure to properly clean a belt can create a build-up of material on the return idlers and thus affect belt training. In addition, it can create an enormous clean-up problem if the material floats to the floor over the length of the conveyor.

The cleaning device is located at the head end of the conveyor and placed so that the scraped material falls back into the discharge chute. In some cases the scrapper is located further back allowing the material to fall into a small hopper and then it is screw conveyed back into the main discharge chute. Rubber and steel blades and brush cleaners are available. The rubber and steel bladed scrapers are the most common and those are usually counterweighted or spring loaded to maintain the necessary pressure against the belt. The brush type cleaners are power driven and are operated at speeds greater than the belt speed for effective cleaning.

Belt Conveyor Sizing Calculations

The capacity of a belt conveyor is determined by the speed and width of the belt and carrying capacity of the idler. The 20 degree troughing idler, shown in Figure 1.7, is commonly used for most materials. The 45 degree troughing idler has greater carrying capacity than the 20 degree idler and is normally used in high capacity applications for handling light materials, such as wood chips, bark and bagasse. The flat idler is seldom used except in feeder applications and in conveyor sections where the use of intermediate discharge plows are required.

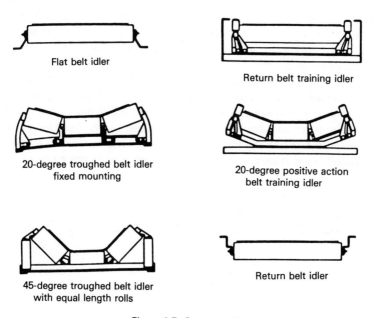

Flat belt idler

Return belt training idler

20-degree troughed belt idler
fixed mounting

20-degree positive action
belt training idler

45-degree troughed belt idler
with equal length rolls

Return belt idler

Figure 1.7. Conveyor idlers.

Assuming that the style idler has been selected for a particular application, the belt size required to satisfy the capacity demands of the conveyor is primarily a function of belt speed.

In arriving at the belt speed it is of prime importance to analyze the material flow at both the loading and discharge ends of the conveyor. The higher the belt speed the more turbulent the condition upon contact of the material with the belt. This is especially true if the material flow is not in the direction of belt travel. Ideally, material velocity should equal belt speed and travel in the same direction. Light and fine materials require belt speeds which would minimize dusting at the loading point. With light and fluffy materials, improper belt speeds can blow the material off the belt. A turbulent atmosphere at the loading point can also degrade fragile materials and with abrasive or sharp lumpy materials it can cause abnormal wear and tear of the belt. In cases where flow to the belt is pulsating or where flow is alternating at two or more loading points, the basis for speed selection may be one which results in an even and continuous load on the belt, especially if uniform discharge is desirable.

In choosing belt speeds, consideration should also be given to the discharge end of the conveyor, especially when handling wet and sticky materials. Speeds should be selected which will result in the material being completely thrown from the belt. However, in this type application, cleaning of the belt is also of importance so that speeds should be moderated to keep wear on the belt scrapers to a minimum as this is a high maintenance but necessary item.

Táble 1.12 gives the maximum recommended belt speeds for different belt widths for different classes of materials. For belts with discharge plows, the maximum belt speed is 200 fpm. For belts where material is visually inspected and manually picked off the belt, speeds range from 50—75 fpm.

Table 1.13 lists the capacities of belt conveyors with 20 degree troughing idlers as a function of belt speed and material density. Capacities are included based on material surcharge angles of 20 degrees and 30 degrees, as shown in Figure 1.8. The surcharge angle along with the belt width, troughing idler and edge distance determines the volume of material. The surcharge angle varies with the material and is usually 10 to 15 degrees less than the angle of repose.

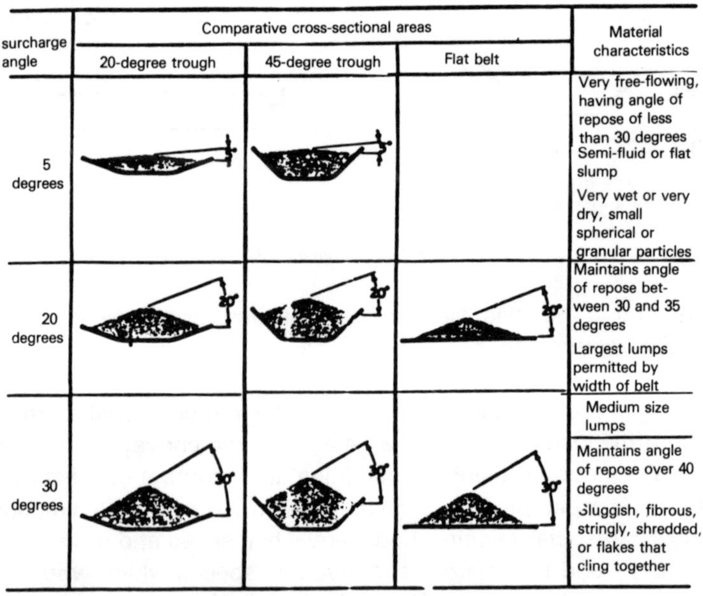

surcharge angle	Comparative cross-sectional areas			Material characteristics
	20-degree trough	45-degree trough	Flat belt	
5 degrees				Very free-flowing, having angle of repose of less than 30 degrees Semi-fluid or flat slump
				Very wet or very dry, small spherical or granular particles
20 degrees				Maintains angle of repose between 30 and 35 degrees
				Largest lumps permitted by width of belt
30 degrees				Medium size lumps
				Maintains angle of repose over 40 degrees
				Sluggish, fibrous, stringly, shredded, or flakes that cling together

Figure 1.8. Belt loading sections.

Table 1.12. Maximum Recommended Belt Speeds Based on Materials

Material Characteristics	Material Example	Maximum Recommended Belt Speed, Feet per Minute Belt Width, inches											
		14	16	18	20	24	30	36	42	48	54	60	72
Half max. lumps, mildly abrasive	Coal, earth	300	300	400	400	500	600	650	700	700	700	700	700
Sized or unsized Very abrasive	Slag, coke, ore, stone, cullet	300	300	400	400	500	600	650	650	650	650	650	650
Flakes	Wood chips, bark, pulp	400	450	450	500	600	700	800	800	800	800	800	800
Granular, 1/8″ to 1/2″ lumps	Grain, coal, cottonseed, sand	400	450	450	500	600	700	800	800	800	800	800	800
Fines light, fluffy, dry, dusty	Soda ash, pulverized coal					150—250 feet per minute							
Fines Heavy	Cement, flue dust					150—300 feet per minute							
Fragile, where degradation is harmful	Coke, coal					100—250 feet per minute.							
Wet Fines	Soap chips					75—200 feet per minute							
	Filter cake					50—100 feet per minute							
Light odd shape	Refuse					100—200 feet per minute							

Table 1.13. Capacity of Belt Conveyors with 20° Troughing Idlers

Capacity, Loading w/20 Deg. Surcharge — Tons per Hour

Material Weight Pounds per Cu. Ft.	Belt Width, Inches	Belt Speed, Ft. per Min.							
		100	200	300	400	500	600	700	800
25	14	8	16	24	32				
	16	10	21	31	42	52			
	18	13	27	40	54	68			
	20	17	33	50	67	84			
	24	25	50	75	100	125	150		
	30	40	81	121	162	202	243	283	
	36	58	117	176	235	293	352	411	470
	42	81	162	243	325	406	487	568	650
	48	110	220	330	440	550	660	770	880
	54	142	285	427	570	712	855	997	1140
	60	180	360	540	720	900	1080	1260	1440
	72	277	555	832	1110	1387	1665	1942	2220
50	14	16	32	48	65				
	16	21	42	63	84	105			
	18	27	54	81	108	135			
	20	34	67	101	135	168			
	24	50	100	150	200	250	300		
	30	81	162	243	324	405	486	567	
	36	117	235	352	470	587	704	822	940
	42	162	325	487	650	812	974	1137	1300
	48	220	440	660	880	1100	1320	1540	1760
	54	285	570	855	1140	1425	1710	1995	2280
	60	360	720	1080	1440	1800	2160	2520	2880
	72	555	1110	1665	2220	2775	3330	3885	4440
100	14	32	64	96	129				
	16	42	84	126	168	210			
	18	54	108	162	216	270			
	20	67	135	202	270	337			
	24	100	200	300	400	500	600		
	30	162	324	486	648	810	972	1134	
	36	235	470	705	940	1175	1410	1645	1880
	42	325	650	975	1300	1625	1950	2275	2600
	48	440	880	1320	1760	2200	2640	3080	3520
	54	570	1140	1710	2280	2850	3420	3990	4560
	60	720	1440	2160	2880	3600	4320	5040	5760
	72	1110	2220	3330	4440	5550	6660	7770	8880

Capacity, Loading w/30 Deg. Surcharge — Tons per Hour

Material Weight Pounds per Cu. Ft.	Belt Width, Inches	Belt Speed, Ft. per Min.							
		100	200	300	400	500	600	700	800
25	14	8	16	24	32				
	16	11	23	34	46	57			
	18	15	31	47	63	78			
	20	19	39	59	79	98			
	24	30	60	90	120	150	180		
	30	48	97	145	194	242	291	340	
	36	73	147	221	295	368	442	516	590
	42	101	202	303	405	506	607	708	810
	48	137	275	412	550	687	825	962	1100
	54	178	357	536	715	893	1072	1250	1430
	60	225	450	675	940	1125	1350	1575	1800
	72	345	690	1035	1380	1725	2070	2415	2760
50	14	16	32	48	65				
	16	23	46	69	92	115			
	18	31	63	94	126	157			
	20	39	79	118	158	197			
	24	60	120	180	240	300	360		
	30	97	194	291	389	485	582	680	
	36	147	295	442	590	737	884	1032	1180
	42	202	405	607	810	1012	1214	1417	1620
	48	275	550	825	1100	1375	1650	1925	2200
	54	357	715	1072	1430	1787	2144	2500	2860
	60	450	900	1350	1880	2250	2700	3150	3600
	72	690	1380	2070	2760	3450	4140	4830	5520
100	14	32	64	96	129				
	16	46	92	138	184	230			
	18	63	126	189	252	315			
	20	79	158	237	316	395			
	24	120	240	360	480	600	720		
	30	194	389	583	778	972	1166	1361	
	36	295	590	885	1180	1475	1770	2065	2360
	42	405	810	1215	1620	2025	2430	2835	3240
	48	550	1100	1650	2200	2750	3300	3850	4400
	54	715	1430	2145	2860	3575	4290	5005	5720
	60	900	1800	2700	3600	4500	5400	6300	7200
	72	1380	2760	4140	5520	6900	8280	9660	11040

ing materials with up to 1 ½ inch lump sizes. In most of the designs the buckets are mounted on chain. Belts are sometimes used when handling very abrasive materials to avoid the wear and maintenance problem that would result from the use of a chain. Belts are also commonly used for economical reasons, where corrosion and sanitary considerations require special and ex- pensive chain materials such as stainless steel and monel.

Operating speeds vary from 200 fpm to 300 fpm, depending on the diameter of the head pulley, and are based on the centrifugal force at the head pulley being about 2/3 of the force of gravity. This relationship is shown by the formula:

$$N = \frac{44.4}{R^{½}} \qquad \text{(Equation 1.14)}$$

where:

    N = Speed of Head Pulley, rpm
    R = Radius to the center of bucket mass, ft.

At these speeds, discharge from the bucket will begin at about a 45 degree angle from the vertical centerline of the pulley. For calculation purposes R can be assumed to be the radius to the center of gravity of the bucket.

Theoretically this results in all of the material being thrown into the discharge spout. However, these calculations are based on material leaving the bucket at zero frictional resistance, and on the bucket being ideally shaped and without the adhesion of product to the inside surfaces of the bucket. In practice this is not the case and the calculated speeds may not result in all of the material being discharged but rather may result in a certain amount being returned down the leg of the elevator into the boot. This phenomenon is known as "back-legging" and it results in an added recirculating load to the elevator. This recirculating load varies based on the material. In the selection of an elevator, the capacity should be increased by at least 15 percent to com- pensate for any "back-legging."

The diameter of the boot pulley should not be less than ¾ of the diameter of the head pulley. A smaller diameter will develop centrifugal forces which are greater than the gravitational forces of the material, resulting in the product being thrown out of the bucket. The most ideal pick-up condition is when the diameter of the foot pulley is equal to the diameter of the head pulley. The capacity of an elevator should be based on the buckets being loaded a max- imum of 60 to 75 percent.

Continuous Discharge

Bucket elevators of this type have buckets which are spaced continuously on a chain or belt. The elevator is designed to handle both fine and lumpy materials up to 4 inches. Since the material is discharged from the bucket by gravity rather than centrifugal action, it is important that the material is dry and free flowing. Continous discharge elevators are also sometimes used for handling fragile or abrasive materials.

Continuous elevators are usually operated at chain or belt speeds of 125 feet per minute. The product is normally loaded directly into the bucket through the use of a loading leg. A small amount of material does bounce off the buckets, as they pass the loading leg, and falls into the boot where it is scooped up by the buckets. The product is discharged by gravity into the discharge spout as the buckets pass down the leg at the head section. The buckets are shaped so that the back of each bucket acts as a chute for the material discharging from the adjacent bucket.

The capacity of the elevator should be based on the buckets being loaded a maximum of 75 percent for fine materials and 60 percent for lumpy materials. The possibility of "back-legging" does exist and a 15 percent capacity correction shoud be used in the selection of an elevator.

In addition to the centrifugal and continuous type elevators as previously described, other elevator designs are available. These include the positive discharge elevator, where the buckets are carried between a double strand of chain and are inverted directly over the discharge spout as the product discharges by gravity. They operate at speeds of 120 fpm and are well suited for handling light, fluffy and sticky materials.

For large capacities of 2000−12,000 cu. feet per hour, the continuous discharge super capacity elevator is available. This is a double chain type elevator, with continuous type buckets carried between the chain, and it operates at approximately 100 feet per minute.

The pivot-type bucket elevator is used to handle free flowing granular materials and is especially suited for handling fragile materials through horizontal and vertical conveying paths.

Bucket Elevator Components

The carrier for the buckets can either be chain or belt and it is usually selected based on the strength, speed, capacity, abrasion and corrosion requirements of the application. Chain elevators are easier to operate since engagement of the cháin to the sprocket wheel teeth results in a very positive transmission of motion and a positive means for tracking the chain. Belts are motivated by the friction between the head pulley and belt and there are no positive means for tracking the belt. The belt must be properly tracked to prevent it from rubbing against the elevator casing. This becomes an especially serious and difficult problem when handling products that might stick to the underside of the belt or head pulley, or build up in the boot section. Because of the tracking problem, take-ups on belt elevators are normally located in the tail section so that the head shaft can be levelled and firmly supported in pillow block bearings.

There are a wide variety chains used for both centrifugal and continuous discharge elevator service. The combination chains and fabricated steel bushed chains are by far the most common. The combination chain consists of a malleable iron center link with steel sidebars. Various special formulations of malleable iron are available from different manufacturers and these offer greater wear resistance for handling abrasive materials. The fabricated welded steel chains are made entirely from steel and are stronger than the combination chains but do not wear as well.

Each type of chain has numerous built-in attachments available for fastening the bucket to the chain. The majority of centrifugal discharge elevators use a Style A or Style A-A malleable iron bucket for either chain or belt applications. The two styles are basically the same. The Style A-A has a thick reinforced lip at the front edge for extra wear protection when handling the more abrasive materials. Special bucket designs are also available for handling sticky and very finely pulverized products.

Buckets for continuous discharge elevators are made of welded steel and come in various configurations to handle an assortment of materials. The 38 degree front bucket is generally used to meet the requirements for handling most materials. Unlike the Style A and A-A buckets, the continuous type bucket is not designed for digging material out of the elevator boot. It usually presents serious maintenance problems when inadvertently used in this manner especially with belt applications.

Belts are also used for both centrifugal and continuous discharge elevators. They are the same belts used for belt conveyors. However, since this belt is fitted with buckets and since the belt operates in a totally enclosed casing, elevator service is recognized to be a more severe application. Consequently, belt selection is not only based on tension considerations, but also on the manufacturer's minimum ply specifications to resist bucket pull-out. The buckets are attached to the belt using large round flat head bolts and at times a bucket may break loose due to the abnormal digging forces generated in the boot.

Slack in the chain or belt is taken up using an external screw take-up, usually located in the boot section. When the chain is expanding and contracting due to temperature, a gravity take-up is sometimes used. In recent years, the internal gravity take-up has received greater recognition, since the entire take-up assembly, including the dry bearing, is totally enclosed in the boot section. This design completely eliminates material leakage and spillage through the shaft seal, and eliminates housecleaning problems usually associated with bucket elevators.

Capacity and Horsepower Requirements

Bucket elevator capacities for standard centrifugal and continuous type elevators with chain are given in Tables 1.14 and 1.15. The listed capacities are based on buckets being filled 75 percent and are proportional to the weight of material being carried and the chain speed. The chain speeds apply for most materials and are essentially based on the forces as previously shown. Light and fluffy materials require 15—20 percent lower chain speeds to lessen the air turbulence created by the fan action of the pulley and buckets. In addition these lower speeds delay the discharge of the material to a position closer to the discharge spout.

The horsepower requirements can be calculated using the formula:

$$HP_T = \frac{T (H + H_o)}{990 \times e} \qquad \text{(Equation 1.15)}$$

where:

HP_T = Elevator Total Horsepower
T = The maximum capacity of the bucket elevator, tons per hour.
H = Vertical Height of Elevator in feet.
H_o = Height factor to overcome boot pulley friction. H_o is 30 for spaced buckets and 10 for continuous buckets.
e = Drive efficiency from Table 1.9 in the Screw Conveyor Section.

The factor for correction of pick-up loss, H_o, is an estimated value. The actual power consumed in the boot section will vary for each material and elevator design and can only be determined from operating data. The maximum capacity of the elevator, T, is based on the buckets loaded 100 percent and it should not be confused with the desired operating capacity of the elevator. Too frequently elevators are overloaded because of uncontrolled and fluctuating feed rates.

VIBRATORY CONVEYOR

Materials are conveyed in vibratory conveyors through use of a spring system which develops a harmonic motion in the pan of the conveyor. The material is moved forward by a series of throws and catches resulting from the upward and forward movement of the pan. In addition to the spring system and pan, other components include a supporting base and an eccentric drive, which imparts a controlled motion to the pan.

Vibratory conveyors are commonly used for handling abrasive and fragile materials, in granular or large sizes and at temperatures from ambient up to 1500 degrees fahrenheit. They are often used to convey materials from the discharge points of crushers and shredders because of their ability to absorb

Table 1.14. Capacity Data for Centrifugal Discharge Elevators

Elevator No.	Buckets Size	Spacing	Chain Speed F.P.M.	Cu.Ft. Hr.	Capacity* TPH Material Wt./Cu.Ft. 35	50	75	100
101	6 × 4	13	225	280	4.9	7.0	10.5	14.0
102	8 × 5	16	230	540	9.5	13.5	20.2	27.0
103	8 × 5	16	230	540	9.5	13.5	20.2	27.0
104	8 × 5	16	260	610	10.7	15.2	22.9	30.5
105	8 × 5	16	260	610	10.7	15.2	22.9	30.5
106	10 × 6	16	230	940	16.4	23.5	35.2	47.0
107	10 × 6	16	230	940	16.4	23.5	35.2	47.0
108	10 × 6	18	268	960	16.9	24.0	36.0	48.0
109	10 × 6	18	268	960	16.9	24.0	36.0	48.0
110	10 × 6	16	260	1050	18.4	26.2	39.4	52.5
111	12 × 7	18	268	1540	26.9	38.5	57.7	77.0
112	12 × 7	16	260	1670	29.2	41.7	62.6	83.5
113	12 × 7	18	306	1740	30.4	43.5	65.2	87.0
114	12 × 7	16	304	1950	34.1	48.7	73.1	97.5
115	14 × 7	19	260	1700	29.8	42.5	63.8	85.0
116	14 × 7	18	268	1850	32.4	46.2	69.4	92.5
117	14 × 7	16	260	2020	35.3	50.5	75.7	101.0
118	14 × 7	19	304	1990	34.8	49.7	74.6	99.5
119	14 × 7	18	306	2110	36.9	52.7	79.1	105.5
120	14 × 7	16	304	2360	41.3	59.0	88.5	118.0
121	16 × 8	19	262	2540	44.4	63.5	95.2	127.0
122	16 × 8	18	248	2540	44.4	63.5	95.2	127.0
123	16 × 8	19	304	2940	51.4	73.5	111.0	147.0
124	16 × 8	18	306	3120	54.6	78.0	117.0	156.0

*Buckets filled to 75% theoretical capacity.

Table 1.15. Capacity Data for Continuous Discharge Elevators

Elevator No.	Buckets Size	Spacing	Chain Speed F.P.M.	Cu. Ft. Per Hr.	Capacity* TPH Material Wt. Per Cu. Ft. 35	50	75	100
701	8 × 5 × 7-3/4	8	125	680	12	17	25	34
702	8 × 5 × 7-3/4	8	125	680	12	17	25	34
703	10 × 5 × 7-3/4	8	125	840	15	21	32	42
704	10 × 5 × 7-3/4	8	125	840	15	21	32	42
705	10 × 7 × 11-5/8	12	125	1080	19	27	41	54
706	10 × 7 × 11-5/8	12	125	1080	19	27	41	54
707	12 × 7 × 11-5/8	12	125	1300	23	32	49	65
708	12 × 7 × 11-5/8	12	125	1300	23	32	49	65
709	14 × 7 × 11-5/8	12	125	1520	26	38	57	76
710	14 × 7 × 11-5/8	12	125	1520	26	38	57	76
711	12 × 8 × 11-5/8	12	125	1560	27	39	58	78
712	12 × 8 × 11-5/8	12	125	1560	27	39	58	78
713	14 × 8 × 11-5/8	12	125	1820	32	45	68	91
714	14 × 8 × 11-5/8	12	125	1820	32	45	68	91
715	16 × 8 × 11-5/8	12-1/8	125	2080	36	52	78	104
716	16 × 8 × 11-5/8	12	125	2080	36	52	78	104
717	18 × 8 × 11-5/8	12-1/8	125	2340	41	58	88	117
718	18 × 8 × 11-5/8	12	125	2340	41	58	88	117

*Based on buckets filled to 75% theoretical capacity.

the high impact loads associated with materials discharging from this equipment. The simple pan construction has no moving parts or pockets and consequently is easily cleaned and is ideal for handling foodstuffs and chemicals where contamination is undesirable. Vibratory conveyors are best suited for handling lumpy and granular materials. Fine materials are handled with more difficulty. Today the vibratory conveyor is considered to be a low maintenance, rugged and dependable machine as illustrated by its wide acceptance and usage in the foundry industry.

Vibratory Conveyor Components

The modern day vibrating conveyor is based on the principal of natural frequency design. In this approach the natural frequency of the vibrating assembly is essentially equal to the frequency of the drive. By operating at the natural frequency of the spring system, the energy required to initially compress each individual spring along the length of the pan is stored in the spring system and released by each individual spring to move the pan forward. Consequently, once the springs are deflected at start-up, the reciprocating or harmonic motion is maintained and the power consumption of the synchronized drive is reduced about 90 percent to the level of the energy required to overcome internal friction. Once the conveyor is started, each spring acts as an individual drive with the motor acting as an energy source for supplying make-up energy to each spring. For a particular spring design, the number of springs in a system is determined by the weight of the pan. The springs are normally equally spaced along the length of the pan and positioned to give a 30 degree stroke angle with the horizontal plane. This design concept allows for the vibration of significant weights and makes possible conveyors with lengths of up to 300 feet or longer.

Vibratory conveyors can produce severe dynamic forces which can be transmitted to the building steel or supporting structure. This has long been recognized as one of the major disadvantages of this conveyor. In recent years designers have addressed this problem and developed satisfactory designs for isolating these forces. The two principal isolation methods are shown in Figure 1.10 and are as follows:

1. Balance Design — In this design, a continuous counter-weight is added which is equal to the vibrating weight of the conveyor and supported from the conveyor base with a duplicate spring system. The counter-weight and pan are driven 180 degrees out-of-phase using a common drive shaft. This design eliminates about 90 percent of the dynamic forces normally developed by the conveyor.
2. Isolated Design — This design features the addition of counterweights to the base of the conveyor plus the use of isolation springs to support the conveyor. This design eliminates about 80 percent of the dynamic forces developed by the conveyor.

The above two methods can also be combined for greater isolation of the dynamic forces from supporting structures. Prior to the above isolation methods, most conveyor applications were limited to ground floor installations where the conveyor could be mounted on a heavy concrete foundation. This design is still commonly used as an acceptable method for absorbing the reaction forces.

Size Determination

The capacity of a vibrating conveyor is a function of the width of the pan, depth of material and the velocity of the material as it moves toward the discharge. The depth of material is usually held to within 2 to 3 inches for granular materials and below 2 inches for fine materials. The velocity of the material is a function of the characteristics of the material and the motion of the pan. In general fine materials do not perform as well as granular materials. For granular, non-sticky type materials conveying speeds of 45 to 90 feet per minute are obtainable. Fine materials of less than 100 mesh will usually have conveying speeds that fall in the 10 to 45 feet per minute range. Materials which are very fine and have a tendency to aerate usually have very low conveying speeds and these materials are usually better handled by other equipment.

The conveyor size, expressed as pan width in inches, can be calculated using the following formula:

$$W = \frac{4800 \times C}{S \times D \times d}$$
(Equation 1.16)

where:

W = Pan Width, inches
C = Capacity, tons per hour
S = Material Conveying Velocity, feet per minute
D = Material Bulk Density, pounds per cu. ft.
d = Bed Depth, inches

Vibratory conveyors are normally available in the following standard pan sizes: 6, 8, 12, 18, 24, 30, 36, 42, 48 and 60 inches. Special designs are also available in one foot increments up to widths of 120 inches.

PNEUMATIC CONVEYOR

In the past three decades pneumatic conveying has grown to where it is now one of the most desirable and widely used conveying techniques in the process industries. It has many advantages over mechanical conveyors and is indeed a conveyor designed to meet the modern day and future requirements of industry.

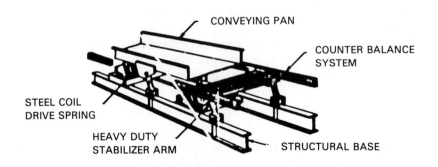

BALANCED HEAVY DUTY CONVEYOR

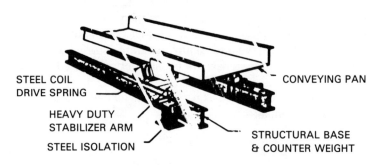

ISOLATED HEAVY DUTY CONVEYOR

Figure 1.10. Isolation methods of vibratory forces.

Pneumatic conveyors use a gas to propel fine and granular solids in a pipeline. The layout of a pneumatic conveying system is relatively simple and usually it will replace the use of a number of mechanical conveyors. Once the feed and discharge points are located, the pneumatic conveyor can be piped to move the product both horizontally and vertically in a single plane or it can be snaked through a number of planes in congested areas. Except for the feed and discharge ends, the run of the conveyor can be described as being low in profile, light in weight and free of any mechanical moving parts. These characteristics give the designer a great deal of flexibility in the layout of a system.

Pneumatic conveyors are considered to be self cleaning conveyors. This is a very desirable characteristic for the food industry and perhaps the most significant reason for its domination of this industry. In addition, there is no leakage or spillage of product along the run of the conveyor and because of its tubular profile, it is considered to be a clean conveyor. These characteristics keep housecleaning to a minimum and are especially of value when handling toxic or potentially explosive products.

Pneumatic conveyors are geared to handle dry fine or granular materials and not the wide variety of materials associated with mechanical conveyors. They are not as efficient as mechanical conveyors and their power requirements are considerably higher. Since the product can travel at speeds up to 5000 fpm, particle degradation is perhaps the biggest disadvantage of conventional pneumatic conveyors. However, in recent years dense phase systems are being designed to handle products at speeds down to 300 fpm to minimize particle degradation. Pneumatic conveyors are also considered to be safer because of the simplicity of design. However, in the event of a pipe joint failure it is possible to pump sizable amounts of material into a room. Normally pipe joints are made using compression type couplings. In recent years, with increased confidence in pneumatic conveying, there has been some movement toward the use of flanged joints.

TYPES OF PNEUMATIC CONVEYORS AND COMPONENTS

There are basically three types of pneumatic conveyors used in the process industries. The three types can be classified based on conveying gas pressures: low pressure, vacuum and high pressure pneumatic conveyors.

Low Pressure Pneumatic Conveyors

The low pressure system as illustrated in Figure 1.11, is a pneumatic conveyor which operates under positive pressure at gas pressures of less than 15 psig. This is a continuous type conveying system in that the product is continuously fed into the system through a rotary airlock feeder and carried by the gas within the conveying line to the discharge point. This system is normally favored over a vacuum type system, especially when delivering material from one feed point to a number of use or storage points. The system is normally designed for a product-air pressure drop across the system of 10 psi, which is well within the discharge pressure limitations of rotary postiive displacement type blowers.

A critical equipment component in this system is the rotary airlock feeder which continuously feeds material into the conveying line. It is designed to feed the product from a low pressure area into the higher pressure pneumatic conveying line with a minimum amount of conveying air leaking past the airlock. An excessive amount of air leakage to atmosphere at this point would reduce the conveying air velocity and perhaps result in an inoperable system. This is one of the reasons why low pressure systems are not recommended for handling abrasive materials. Abrasive materials would rapidly wear the airlock rotor and housing and thus allow an abnormal amount of air leakage at this

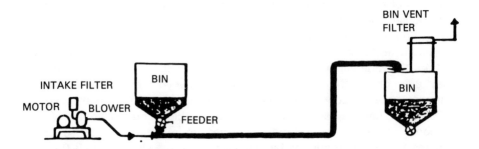

Figure 1.11. Low pressure pneumatic conveyor.

point. Normally the clearance between the rotor and housing is 0.003 to 0.005 inches for six or eight blade rotors. There will always be some air leakage through this gap and it is usually referred to as by-pass air. By-pass air leakage, Fa, varies with the airlock size, rotor clearance and the differential pressure across the airlock. It can be as high as 20—60 cfm for a 0.4 cu. ft. displacement valve operating at 8 psig line pressure. This is a significant quantity of air for systems of 3 inch diameter and smaller.

In addition to by-pass air leakage there is also air leakage due to the rotor pocket being pressurized as it discharges material into the pneumatic line. As this pocket of air rotates to the low pressure side of the airlock, the air is released. This leakage is normally referred to as blow-back air and can be calculated using the formula:

$$F_b = Va \times Da \qquad \text{(Equation 1.17)}$$

where:

F_b = Blow-Back air, cfm
Va   = Airlock Speed, rpm
Da   = Airlock Displacement, air cu. ft. per rev.

The pressurized gas or air for a pressure type pneumatic conveying system is supplied using a rotary positive displacement blower. These blowers are available from a number of manufacturers and are available with pressure ratings up to 18 psig. An important characteristic of this type of blower is that it automatically operates at a pressure level equal to resistance pressure of the system. For any blower speed a relatively constant volume of air is provided over a full range of discharge pressures. These characteristics are ideal for pneumatic conveying since they compensate for variations in material flow or any upset condition that may occur. In the event of the formation of a material plug in the conveying line, the blower discharge pressure will automatically increase to the level required to maintain flow.

The blower discharge air temperature may be calculated using the formula:

$$\Delta t = \frac{T_1 \times BHP \times 0.92}{.0153 \times P_1 \times V_1}$$

(Equation 1.18)

where:

Δt = Temperature Rise, $^\circ$F
T_1 = Inlet Air Temperature, $^\circ$R
BHP = Brake Horsepower
P_1 = Inlet Air Pressure, psia
V_1 = Inlet Air Flow, cfm.

Some products such as milk powder, PVC, etc., are temperature sensitive and may require the use of an aftercooler at the blower discharge. Aftercoolers are also used in closed loop systems to keep the conveying gas below the temperature limitations of the blower. In a closed loop system the conveying gas, which is usually air or nitrogen, is recycled back to the blower and re-compressed.

In most applications, the mixture of solids and air is separated at the terminal end of the system using a high efficiency dust collector. National air quality standards for particulate discharge into the atmosphere have essentially eliminated the use of pneumatic cyclones as final separators except for handling non-dusty products such as plastic pellets.

The reverse air type dust collector is normally used in pneumatic conveying systems. With this type collector the dirty gas is filtered through the outer surface of tubular shaped filter bags which are supported by wire cages or perforated metal tubes. The dust cake is cleaned off by a reverse flow of compressed air at controlled intervals. Usually a row or a number of bags are cleaned simultaneously. The short blast of compressed air, which can last for a period of .2 to 1.0 seconds, sends a shock wave down the bag, flexing the bags which frees the deposited dust. The dust drops into the hopper of the dust collector and continuously discharges through a rotary airlock to downstream equipment. During the cleaning cycle the reverse flow of air momentarily interrupts the flow of dirty air through the bags being cleaned. Since only a few bags are cleaned at one time and an extremely short time is taken for cleaning, the result is an uninterrupted flow of air into the collector.

The reverse air type collectors operate at a higher air to cloth ratio than the mechanical shake type cleaning units. Since they require less cloth area for a given application, the units are smaller in size than the mechanical shake type collectors and easier to fit into the process stream. Air to cloth ratios are held to between 5:1 and 7:1 for most pneumatic conveying applications versus 10:1 for conventional dust collector applications. Air to cloth ratio is defined as CFM of air flow to square feet of filter cloth area.

Vacuum Pneumatic Conveyors

The vacuum type pneumatic conveyor shown in Figure 1.12 operates under negative pressure and is limited to conveying pressures of 12 inches of mercury. Material enters the system through a slide gate or rotary feeder and is conveyed under negative pressure to a dust collector where the material is separated from the air. The product is discharged from the collector using a rotary airlock. The air is pulled through the line using a rotary positive displacement blower located downstream from the dust collector. It is a continuous type system and normally used when conveying is required from a number of storage or process points to one discharge point. Other applications include railroad car unloading and the transfer of product from bag dump stations.

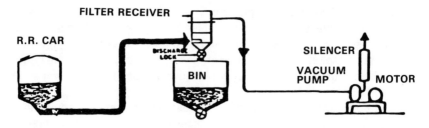

Figure 1.12. Vacuum pneumatic conveyor.

The equipment components for a vacuum system are basically the same as those required for a pressure system. In this application the dust collector is designed for operation under negative pressure. The rotary positive displacement blowers as previously described are also used for vacuum service and are limited to pressure levels of 12 to 15 inches of mercury. A small secondary filter is normally used after the dust collector to protect the blower in the event of bag failure. A high level switch is almost always included in the dust collector hopper to detect product build-up due to a malfunction of the rotary airlock. Here again excessive wear of the rotary airlock can cause failure of the pneumatic system since air will be short circuited into the system at this point. Vacuum type systems are not recommended for handling abrasive products.

Vacuum-Pressure Pneumatic Conveyors

The vacuum-pressure type pneumatic conveyor as shown in Figure 1.13 gives the designer the opportunity to combine the advantages of both the vacuum and pressure type pneumatic conveyors. Originally the design concept was to use one blower to develop the vacuum and pressure requirements of the system. However, this meant that the compression requirements of the entire system had to remain within the compressibility factor of the blower. In reality this was difficult to design since occasional problems on the pressure

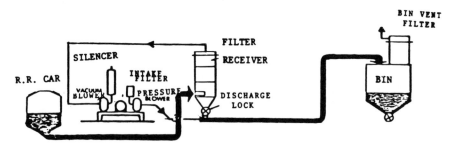

Figure 1.13. Vacuum-pressure pneumatic conveyor.

side of the system would create greater pressure requirements, which would in turn have adverse affects on the vacuum side of the system and vice versa. For this reason, today most systems are designed with two blowers, one pressure and one vacuum, usually V-belt driven from a common motor.

In a typical system the product is drawn into the vacuum side of the system and conveyed to the dust collector. Usually the vacuum run is made as short as possible. The product is separated in the dust collector and discharged by use of a rotary airlock into the pressure leg of the system. The airlock is a particularly critical item in this application, since an excessive amount of air leakage can upset both the vacuum and pressure systems. The product is then conveyed by pressure to one or a number of use points. The vacuum-pressure pneumatic system is commonly used for unloading railroad cars into a number of storage silos.

High Pressure Pneumatic Conveyors

The high pressure or dense phase pneumatic conveyor, illustrated in Figure 1.14, operates at gas pressures between 15 and 100 psig. It is a batch type conveyor and because of its low conveying velocities of 300 to 2000 feet per minute, it is best suited for handling abrasive materials and materials where particle degradation is undesirable. Unlike dilute phase conveyors, the material occupies a significant portion of the cross sectional area of the conveying line and is actually pushed through the system.

The dense phase system uses a pressure vessel or blow pot to contain and feed the material into the conveying line. The vessel is first loaded with material under atmospheric pressure through the inlet valve. It is then sealed and pressurized up to the operating pressure using the conveying gas. The outlet valve is then opened and the material is pushed through the conveying line by the high pressure gas at very high material to air ratios. Recent designs include the use of booster stations along the conveying line for injecting additional high pressure gas into the line to maintain flow. A compressor is normally used to supply the high pressure gas or the gas can be supplied from a central plant source.

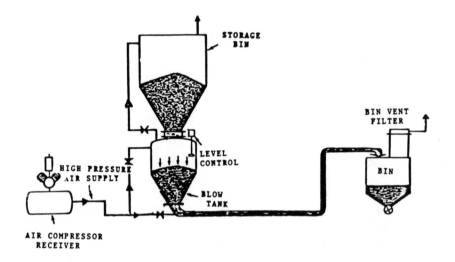

Figure 1.14. High pressure pneumatic conveyor.

Size Determination

The line size and blower requirements of a pneumatic conveying system can be arrived at by selecting a conveyor air or gas velocity which will propel the solids through the pipeline and by determining the energy requirements of the conveying air. The energy requirements are normally expressed in pressure units and are maintained within the compression requirements of the rotary blowers. For low pressure type systems it is desirable to maintain a system pressure drop of less than 10 psi. For vacuum type systems pressure drop requirements of less than 6 psi or approximately 12 inches of mercury is the usual practice. Unfortunately, there are no absolute theoretical methods for calculating conveying velocities or pressure drops. Consequently, all pneumatic conveying systems are sized from either experience, test data or empirical equations or from a combination of any of these approaches.

Equations are not available for calculating the conveying air velocity. The velocity required to float or hold a particle in suspension can be calculated from Stokes' law. However, the conveying velocity is greater than the floating velocity and it can only be determined from actual experience with the subject material or test work. In general, powdered materials can be conveyed in low pressure or vacuum systems with maximum line velocities or terminal velocities ranging from 4000 to 5500 fpm. Granular and pelletized materials hop through the line and require higher velocities ranging from 5500 to 7500 fpm. These velocity recommendations are not applicable for systems greater than 6 inches in diameter or greater than 500 feet in overall length. The actual material velocity in the line can be as low as 40 percent of the conveying gas velocity for pelletized materials and as high as 90 percent for powdered materials.

The total energy required to convey material in a pipeline using low pressure systems can be arrived at by calculating, with empirical equations, the energy usage due to the following characteristics:

A. The energy required to accelerate the product from zero velocity at the feed point to the conveying air velocity. This can be represented by the equation:

$$E_1 = \frac{Mv^2}{2g} \qquad \text{(Equation 1.19)}$$

where:

E_1 = Acceleration Energy, ft.-lbs./minute
M = Conveying Rate, lbs./minute
v = Max. Conveying Gas Velocity, ft./sec.
g = Acceleration due to gravity, ft./sec.^{2}

B. The energy required to convey the product in the horizontal section of pipe. This can be approximated using the formula:

$$E_2 = Md_1f \qquad \text{(Equation 1.20)}$$

where:

E_2 = Horizontal Energy, ft.-lbs./min.
d_1 = Horizontal Distance, ft.
f = Coefficient of Friction of the Material, Table 1.16

The coefficient of friction is the tangent of the angle of slide between the material being conveyed and the pipeline material.

C. The energy required to convey the product in the vertical section of pipe. This can be calculated from the simplified equation:

$$E_3 = Md_2 \qquad \text{(Equation 1.21)}$$

where:

E_3 = Vertical Energy, ft.-lbs./min.
d_2 = Vertical Distance, ft.

Table 1.16. Coefficient of Friction of Bulk Materials

Material	Bulk Density Lbs./Cu. Ft.	Coefficient of Friction
Alumina, Fine	60	0.7
Aluminum Hydrate, Fine	18	0.7
Aluminum Sulfate, Granular	54	0.6
Bauxite, Granular	70	0.5
Bone Char, Granular	40	0.7
Borax, Powdered	53	0.9
Calcium Bicarbonate, Granular	62	0.5
Calcium Carbide, Powdered	40	1.0
Calcium Oxide, Powdered	27	1.0
Carbon, Ground	50	0.4
Carbon Coke, Granular	30	0.6
Cellulose Acetate, granular	10	0.7
Cement, Portland	75-85	0.9
Charcoal, Wood, Granular	26	0.7
Chocolate, Powdered	40	1.0
Chromic Acid, Flake	75	0.5
Clay, Powdered	80	1.0
Coal, Bituminous, Loose	50	0.7
Cocoa, Beans	30-40	0.5
Cocoa, Powdered	30-35	0.7
Coffee Beans, Green	32	0.5
Coffee Beans, Roasted	22-26	0.5
Copper Oxide, Powder	190	0.9
Corn, Grits	40	0.5
Dolomite, Granular	46	0.9
Gypsum, Powdered	60	1.0
Iron Oxide, Pigment	25	0.9
Kaolin, Fine	22	1.0
Lime, Pebble	56	0.6
Lime, Hydrated, Powder	40	0.9
Limestone, Pulverized	85	0.9
Magnesium Carbonate, Powder	9	0.9
Milk, Powder	20	1.0
Oats, Rolled	18	0.6
Phosphate, Super, Ground	51	1.0
Phthalic Anhydride, Flake	42	0.5
Salt, Granulated	80	0.6
Sand, Coarse Sized	90-100	0.6
Sawdust, Dry	10-30	0.8
Soda Ash, Light	20-35	0.8
Soda Ash, Dense	55-65	0.7
Soy Beans, Crushed	34	0.7
Starch, Powder	25-45	1.0
Sugar, Granulated	50-55	0.7
Sugar, Powder	45	1.0
Titanium Dioxide, Powder	25	1.0
Wheat	48	0.5
Zinc Sulfate, Powder	72	1.0

D. The energy required to convey the product through bends in the line. This can be calculated from the empirical equation:

$$E_4 = \frac{Mv^2nd_3f}{gR}$$

(Equation 1.22)

where:

E_4 = Bend Energy, ft.-lbs./min.

n = number of equivalent 90 degree bends the product passes through.

R = Radius of Bend, ft.

d_3 = Bend Length, ft.

The above energy losses are termed product losses. In addition to these losses, there are also air losses due to the clean air flowing through the pipeline and terminal equipment. The pipeline losses can be taken from Table 1.17 and are expressed in inches of water. A pressure loss of 3 inches of water can be assumed for cyclone losses and 6 inches of water for most dust collector losses.

Table 1.17. Friction Loss of Air in Pipes

Air Velocity Ft./Min.	Friction Loss per 100 Ft. of Pipe, Inches Water Pipe Size						
	2	3	4	5	6	8	10
3500	12.0	7.8	5.4	4.2	3.4	2.4	1.9
4000	16.0	9.7	6.8	5.5	4.3	3.2	2.4
4500	20.0	12.0	8.7	6.7	5.3	3.8	3.0
5000	24.0	15.0	11.0	8.5	6.6	4.8	3.7
5500	28.0	18.0	13.0	10.0	8.1	5.6	4.3
6000	33.0	21.0	15.0	12.0	9.2	6.4	5.0
6500	37.0	24.0	17.0	13.5	11.0	7.6	5.8
7000	42.0	27.0	20.0	16.0	13.0	9.0	6.8
7500	48.0	31.0	23.0	18.0	14.0	10.0	7.9
8000	55.0	35.0	25.0	21.0	16.0	12.0	9.0

The following examples will illustrate this method of calculation:

Example A

Design a pneumatic conveying system to convey hydrated lime from the discharge of a storage silo to two use points. The conveying rate is 12,000 PPH or 200 PPM and the conveying distance to the furthest use point is 150 feet horizontal, 25 feet vertical, including three 90 degree and one 45 degree elbows with a bend radius of 2.5 feet. This is a typical pressure type system. Since hydrated lime is a fine powder with a bulk density of about 40 pounds per cu. ft., assume an air conveying velocity of 5500 fpm or 91.7 ft. per sec. From Table 1.16 the coefficient of friction is 0.9.

Product Losses

$$E_1 = \frac{200 \times 91.7^2}{2 \times 32.2} = 26,115 \text{ ft.-lbs./min.}$$

$$E_2 = 200 \times 150 \times .9 = 27,000 \text{ ft.-lbs./min.}$$

$$E_3 = 200 \times 25 = 5000 \text{ ft.-lbs./min.}$$

$$E_4 = \frac{200 \times 91.7^2 \times 3.5 \times 3.9 \times 0.9}{32.2 \times 2.5} = 256,654 \text{ ft.-lbs./min.}$$

The total product losses are:

$$E_T = E_1 + E_2 + E_3 + E_4 = 314,769 \text{ ft.-lbs./min.}$$

Assume that a 3″ dia. sch. 40 pipe will be required for this system. Then the maximum air flow based on a velocity of 5500 fpm is 282 CFM. The product energy usage can be converted to pressures losses in inches of water using the formula:

$$\Delta P_p = \frac{E_4}{5.2 \times F}$$

where:

ΔP_p = Product Pressure Loss, inches of water.
F = Max. Gas Flow, CFM.

therefore

$$\Delta P_p = \frac{314,769}{5.2 \times 282} = 215 \text{ inches } H_2O$$

Air Losses

From Table 1.17, the pressure loss due to the flow of clean air flowing through the horizontal and vertical sections of pipe is 18 inches per 100 ft. of pipe or $18.0 \times 1.75 = 32$ inches of H_2O. The pressure drop across the filter at the terminating end of the system can be assumed at 6 inches of water. Therefore, the total air losses are 38 inches of H_2O.

The pressure drop across the system is the sum of the product losses and air losses or 253 inches of H_2O or 9.1 psi.

Therefore the assumed 3″ dia. line is a good selection for this system. A 2 inch dia. line would result in a total pressure loss far in excess of 10 psi and a 4 inch dia. line would be less economical since the pressure loss would be substantially below 9.1 psi. The rotary positive displacement blower would be rated for an inlet flow of 282 CFM plus the air inleakage thru the airlock and a discharge pressure of 9.1 psig.

Example B

Design a pneumatic conveying system to convey hydrated lime from the discharge of a silo or bag dump station to one use point. The conveying rate is 4,000 PPH or 66.7 PPM and the conveying distance is 75 feet horizontal, 20 feet vertical, including two 90 degree and one 30 degree elbows with a bend radius of 2.5 feet.

This is a typical vacuum type system. As in the above example, assume an air conveying velocity of 5500 fpm or 91.7 ft. per sec. and a coefficient of friction of 0.9.

Product Losses:

$$E_1 = \frac{66.7 \times 91.7^2}{2 \times 32.2} = 8709 \text{ ft.-lbs./min.}$$

$$E_2 = 66.7 \times 75 \times 0.9 = 4502 \text{ ft.-lbs./min.}$$

$$E_3 = 66.7 \times 2 = 1334$$

$$E_4 = \frac{66.7 \times 91.7^2 \times 2.33 \times 3.9 \times 0.9}{32.2 \times 2.5} = 56,981$$

$$E_T = E_1 + E_2 + E_3 + E_4 = 71,526 \text{ ft.-lbs./min.}$$

Assume that a 2″ dia. sch. 40 pipe will be required for this system. Then the maximum air flow based on a velocity of 5500 fpm is 128 CFM.

$$\Delta P_p = \frac{71,526}{5.2 \times 128} = 107 \text{ inches of } H_2O$$

Air Losses

From Table 1.17, the pressure loss due to the flow of clean air flowing through the horizontal and vertical sections of pipe is 28.0 inches of H_2O per 100 feet of pipe or $28.0 \times .95 = 27$ inches of H_2O. The pressure drop across the filter at the terminating end of the system is 6 inches of water. Therefore the total air losses are 34 inches of H_2O.

The pressure drop across the system is the sum of the product losses and air losses or 141 inches of water or 10.4 inches of Hg.

A 2 inch diameter system with a blower rated for 128 inlet CFM plus the air in leakage through the airlock and 10.4 inches of Hg vacuum can be used for this application. Some designers would consider this a marginal system and would prefer a system based on a 3 inch dia. line so that the pressure loss is less than 10 inches of Hg. This is an example where experience with pneumatic conveying may be the determining factor in the selection of a system.

BIBLIOGRAPHY

1. Carman Industries, Inc., Bulletin No. 700, "Vibrating Conveyors."
2. CEMA, Book No. 550, "Classification and Definition of Bulk Materials," CEMA, Washington, D.C.
3. CEMA, Book No. 350, "Screw Conveyors," CEMA, Washington, D.C.
4. CEMA, Book 1966, "Belt Conveyors for Bulk Materials."
5. Fischer, J., "Practical Pneumatic Conveyor Design," Chemical Engineering, June 2, 1958.
6. Goodyear Tire & Rubber Co., Inc., "Handbook of Belting."
7. Hetzel & Albright, "Belt Conveyors and Belt Elevators," John Wiley & Sons, Inc., New York.
8. Link-Belt Div., FMC, "Catalog No. 1000."
9. Meyer Machine Co., Bulletin No. 910-H, "SIMPLEX Conveying Elevators."
10. Stephens-Adamson Mfg. Co., "Catalog No. 66."
11. Wallace Mfg. Co., Bulletin "Industrial Service Bucket Elevators."

CHAPTER 2

CENTRIFUGAL PUMPS

J. R. BIRK AND R. E. SYSKA
The Duriron Company, Inc.
Dayton, Ohio

Pumps for the chemical process industries differ from those used in other industries primarily in the materials from which they are made.

While cast iron, ductile iron, carbon steel, and aluminum or copper-base alloys will handle a few chemical solutions, most chemical pumps are made of stainless steel, nickel-base alloys, or more exotic metals such as titanium and zirconium. Pumps are also available in carbon, glass, porcelain, rubber, lead and whole families of plastics, including phenolics, epoxies and fluorocarbons.

Each of these materials has been incorporated into pump designs for just one reason: to eliminate or reduce the destructive effect of the chemical on the pump parts.

Since the type of corrosive liquid will determine which of these materials will be most suitable, a careful analysis of the chemical to be handled must first be made. Like other types of chemical process equipment, pumps will experience only eight forms of corrosion.

1. General, or uniform, corrosion is the most common type, characterized by essentially the same rate of deterioration over the entire wetted for exposed surface.
2. Concentration-cell, or crevice, corrosion is a localized form resulting from small quantities of stagnant solution in areas such as threads, gasket surfaces, holes, crevices, surface deposits, and under bolt and rivet heads. Usually this form of corrosion does not occur in chemical pumps except perhaps under gaskets, or in designs where the factors known to contribute to concentration-cell corrosion have been ignored.
3. Pitting corrosion is the most insidious, destructive form of corrosion, and very difficult to predict. It is extremely localized. Chlorides in particular are notorious for inducing pitting that can occur in practically all types of equipment.
4. Stress-corrosion cracking is localized failure caused by the combination of tensile stresses and a specific medium. Fortunately, castings due to their basic overdesign, seldom experience stress-corrosion cracking.
5. Intergranular corrosion is a selective form of corrosion at the alloy grain boundaries. It is associated primarily with stainless steels but can also occur with other alloy systems. Unless other alloy adjustments are made, this form of corrosion can be prevented only by heat treating.

6. Galvanic corrosion occurs when dissimilar metals are in contact, or otherwise electrically connected, in a corrosive medium. When it is found necessary to have two dissimilar metals in contact, caution should be exercised to make certain that the total surface area of the less resistant metal far exceeds that of the more corrosion resistant material. This form of corrosion is not common in chemical pumps, but may be of some concern with accessory items that may be in contact with the pump parts, and are subjected to the pumped solution.

7. Erosion-corrosion is characterized by accelerated attack resulting from the combination of corrosion and mechanical wear. The ideal material to avoid erosion-corrosion in pumps would possess the characteristics of corrosion resistance, strength, ductility and high hardness. Few materials possess such a combination. Cavitation is considered a special form of erosion-corrosion that results from the collapse or implosion of gas bubbles against the metal surface in high-pressure regions.

8. Selective-leaching corrosion involves removal of one element from a solid alloy in a corrosive medium. This form of attack is not common to chemical pumps, because the alloys in which it occurs are not commonly used in heavy chemical applications.

CORROSIVES & MATERIALS

Materials for pump applications can, in general, be divided into two very broad categories; metallic and nonmetallic. The metallic category can be further subdivided into ferrous and nonferrous alloys, both of which have extensive application in the chemical industry. The nonmetallics can be further subdivided into natural and synthetic rubbers, plastics, ceramics, glass, carbon and graphite.

For a given application, a thorough evaluation of not only the solution characteristics but also the materials available should be made to ensure the most economical selection.

TYPICAL MATERIALS OF CONSTRUCTION

The most widely used metallic materials of construction for chemical pumps are the stainless steels. Of the many available, the most popular are the austenitic grades, such as the cast equivalents of Type 304 and Type 316, which possess superior corrosion properties compared to the martensitic or ferritic grades.

The stainless steels are used for a wide range of corrosive solutions. They are suitable for most mineral acids at moderate temperatures and concentrations. The notable exceptions are hydrochloric and hydrofluoric acids. In general, the stainless steels are more suitable for oxidizing than for reducing environments. Organic acids and neutral-to-alkaline salt solutions are also handled by stainless steel pumps.

Carbon steel, cast iron, and ductile cast iron are also frequently used for the many mildly corrosive applications found in most plants.

For the more severe or critical services, the high-alloy stainless steels such as Alloy 20 are frequently specified.

Nickel-base alloys, because of their relatively high cost, are generally used only where no iron-base alloy is suitable. This family of corrosion resistant materials includes: pure nickel, nickel-copper, nickel-chromium, nickel-molybdenum, and nickel-chromium-molybdenum alloys.

Aluminum, titanium and copper-base alloys such as bronze or brass are the most frequently used nonferrous metals for chemical pumps. Zirconium has also found application in a few very special areas.

Both natural rubber and synthetic rubber linings are used extensively for abrasive and/or corrosive applications. Soft natural rubber generally has the best abrasion resistance, but cannot be used at as high a temperature as semi-hard natural rubber or the synthetic rubbers such as Neoprene and butyl. In most cases, the hard rubbers and synthetic rubbers also possess better chemical resistance.

Plastics are among the fastest growing families of pump materials. For the ultimate in chemical resistance, the fluorocarbon resins such as polytetrafluoroethylene (PTFE) and fluorinated ethylenepropylene (FEP) are finding wide application. Where strength and chemical resistance are needed, a variety of fiber-reinforced plastics (FRP) are available. Epoxy, polyester, and phenolic are three of the more popular FRP materials. Polyvinyl chloride, polyethylene and polypropylene are also finding application. Plastics are gaining in popularity because they offer the corrosion resistance of the more expensive metals at a fraction of the cost. However, they have strength limitations and it is doubtful that plastics will ever completely replace metals.

For many extremely corrosive services at elevated temperatures, glass or ceramic are the most suitable materials because of their extreme chemical inertness. However, these materials are usually avoided because of their poor mechanical properties.

Carbon or graphite construction is generally used for the same kinds of services as are ceramic or glass. The primary reason for using carbon or graphite instead of glass or ceramic is that the former are suitable for services where HF or alkalis are handled.

Of the several factors that determine whether or not a certain material can be used for a particular pump design, mechanical properties are the most important. Materials may possess outstanding corrosion resistance but may be completely impossible to produce in the form of a chemical pump because of limited mechanical properties. Hence, awareness of these properties is essential for any material being considered in a corrosion evaluation program. Since most materials are covered by ASTM or other specifications, such sources can be used for reference purposes. A table of mechanical properties and other characteristics of proprietary materials not included in any standard specifications should be readily available from the manufacturer of the material.

Weldments or welded construction should impose no limitation, providing the weldment is as good as, or better than, the base material. Materials requiring heat treatment in order to achieve maximum corrosion resistance must be heat treated after a welding operation, or other adjustments must be made, to make certain that corrosion resistance has not been sacrified.

Wall sections in pumps are generally increased over the mechanical design requirements, so that full pumping capability will be maintained even after the loss of some material to the corrosive environment. Parts that are subject to corrosion from two or three sides, such as impellers, must be made considerably heavier than their counterparts in water or oil pumps. Pressure containing parts are also made thicker so they will remain serviceable after a specified amount of corrosive deterioration. Areas subject to high velocities, such as the cutwater of a centrifugal casing, are further reinforced to allow for the accelerated corrosion caused by high velocities in the liquid.

Threaded construction of any type within the wetted parts must be avoided whenever possible. The thin thread is subject to attack from two sides, and a small amount of corrosive deterioration will eliminate the holding power of the threaded joint. Pipe threads are also to be avoided because of their susceptibility to attack.

Gasket materials must be selected to resist the chemical being handled. Compressed asbestos, lead, and certain synthetic rubbers have been used extensively for corrosion services. In recent years, the fluorocarbon resins have come into widespread use, due to their almost universal corrosion resistance.

The power end of pumps consists of the bearing housing, bearings, oil or grease seals, and the bearing lubrication system. This assembly is normally made of iron or steel components, and thus must be designed to withstand the severe environment of the chemical plant. For example, when venting of the bearing housing is required, special means of preventing the entrance of water, chemical fumes, or dirt must be incorporated into the vent designs.

The bearing that controls axial movement of the shaft is usually selected to limit movement to 0.002 in. or less. End-play values above this limit have been found detrimental to mechanical-seal operation.

Water jacketing of the bearing housing may be necessary under certain conditions to maintain bearing temperatures below 225°F, the upper limit for standard bearings.

STUFFING BOX DESIGN

The area around the stuffing box probably causes more failures of chemical pumps than all other parts combined. The problem of establishing a seal between a rotating shaft and the stationary pump parts is one of the most intricate and vexing problems facing the pump designer.

Packings of braided asbestos, lead, fluorocarbon resins, aluminum, graphite and many other materials, or combinations of these materials, have been used to establish the seal. Inconsistent as it seems, a small amount of liquid must be allowed to seep through the packing to lubricate the surface between packing and the shaft. This leakage rate is hard to control, and the usual result is overtightening of the packing to stop the leak. The unfortunate consequence is the rapid scoring of the shaft surface, making it much harder to adjust the packing to the proper compression. Recommendations as to the type of packing to be used for various chemical services should come from the packing manufacturer.

Mechanical shaft seals are used extensively on chemical pumps. Once again, the primary consideration is selection of the proper materials for the type of corrosive being pumped. Stainless steels, ceramics, graphite and fluorocarbon resins are used to make the bulk of the seal parts. Several large manufacturers of this equipment have very complete files on seal designs for various chemical services. Typical seal installations are shown in Figures 2.1 and 2.2.

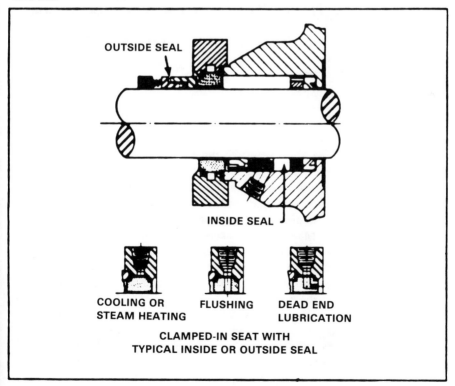

· *Figure 2.1.*

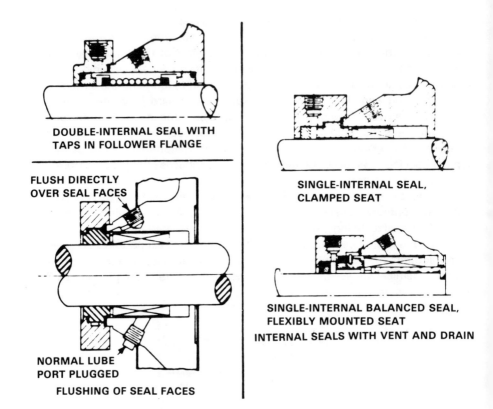

DOUBLE-INTERNAL SEAL WITH
TAPS IN FOLLOWER FLANGE

FLUSH DIRECTLY
OVER SEAL FACES

SINGLE-INTERNAL SEAL,
CLAMPED SEAT

SINGLE-INTERNAL BALANCED SEAL,
FLEXIBLY MOUNTED SEAT
INTERNAL SEALS WITH VENT AND DRAIN

NORMAL LUBE
PORT PLUGGED
FLUSHING OF SEAL FACES

Figure 2.2.

The operating temperature is one of the most important factors affecting the stuffing box sealing medium. Most packings are impregnated with grease for lubrication, but these lubricants break down at temperatures above 250°F. resulting in further temperature increases because of friction. One of the less obvious results of this temperature increase is corrosive attack on the pump parts in the heat zone. Many materials selected for the pumping temperature will be completely unsuitable in the presence of the corrosive at elevated stuffing box temperatures. Another source of heat is the chemical solution itself. These liquids are often in the 300°F. range, and some go as high as 700°F.

The best answer to the heat problem is removal of the heat by means of a water jacket around the stuffing box. While heat conductivity is rather low for most chemical pump materials, the stuffing box area generally can be maintained in the 200°F. to 250°F. range. This cooling is of further benefit in that it prevents the transfer of heat along the shaft to the bearing housing, thereby, eliminating other problems around the bearings.

Stuffing box pressure varies with suction pressure, impeller design, and the degree of maintenance of close-fitting seal rings. Variations in impeller design would include those using vertical or horizontal seal rings in combination with balance ports, as opposed to those using back vanes or pump-out vanes. All impeller designs depend upon a close-running clearance between the impeller and the stationary pump parts. This clearance must be kept as small as possible to prevent excessive recirculation of the liquid, and resulting loss of efficiency. Unfortunately, most chemical pump materials tend to seize when subjected to rubbing contact. Therefore, running clearances must be increased considerably above those for other industries.

At pressures above 100 psig, packing is generally unsatisfactory unless the stuffing box is very deep, and the operator is especially adept at maintaining the proper gland pressure on the packing. Mechanical seals incorporating a balance feature to relieve the high pressure are the best means of sealing at pressures above 100 psig.

In the stuffing box region, the shaft surface must have corrosion resistance at least equal to, and preferably better than that of wetted parts of the pump. In addition, this surface must be hard enough to resist the tendency to wear under the packing or mechanical seal parts. Further, it must be capable of withstanding the sudden temperature changes often encountered.

Since it is not economically feasible to make the entire pump shaft of stainless alloys, and physically impossible to make functional carbon, glass, or plastic shafts, chemical pumps often have carbon steel shafts with a protective coating or sleeve over the steel in the stuffing box area. Cylindrical sleeves are sometimes made so that they may be removed and replaced when they become worn. Other designs use sleeves that are permanently bonded to the shaft to minimize runout and deflection.

Composite shafts using carbon steel for the power end and a higher alloy for the wet end have been used extensively where the high alloy end has acceptable resistance. Since the two ends are joined by various welding techniques, the combination of metals is limited to those that can be easily welded. On such assemblies, the weld joint and the heat affected zone must be outside the wetted area of the shaft.

DESIGNING WITH SPECIAL MATERIALS

A number of low-mechanical strength materials have been used extensively in chemical pump construction. While breakage problems are inherently associated with these materials, their excellent corrosion resistance has allowed them to remain competitive with higher strength alloys. Of course, their low tensile strength and brittleness make them sensitive to tensile or bending stresses, requiring special pump designs. The parts are held together by outside clamping means, and braced to prevent bending. The unit must also be protected from sudden temperature changes and from mechanical impact from outside sources.

Although produced by very few manufacturers, high silicon iron is the most universally corrosion resistant metallic material available at an economic price. This resistance, coupled with a hardness of approximately Brinell 520, provides an excellent material for handling abrasive chemical slurries. The material's hardness, however, precludes normal machining operations, and the parts must be designed for machine grinding. The hardness also eliminates the possibility of using drilled or tapped holes for connecting piping to the pump parts. Therefore, special designs are required for process, piping, stuffing box lubrication, and drain connections.

Ceramics and glass are similar to high silicon iron in regard to hardness, brittleness and susceptibility to thermal or mechanical shock. Pump designs must, therefore, incorporate the same special considerations.

Glass linings or coats on iron or steel parts are sometimes used to eliminate some of the undesirable characteristics of solid glass. While this usage provides for connecting process piping, the dissimilar expansion characteristics of the two materials generate small cracks in the glass, allowing corrosive attack.

Thermosetting and thermoplastic materials are used extensively in services where chlorides are present. Their primary disadvantage is loss in strength at higher pumping temperatures. Phenolic and epoxy parts are subject to gradual loss of dimensional integrity because of their creep characteristics. The low tensile strength of the unfilled resins again dictates a design that will place these parts in compression, and eliminate bending stresses. Typical construction details are shown in Figure 2.3.

Polytetrafluoroethylene and hexafluoropropylene possess excellent corrosion resistance. These resins have been used for gaskets, packing, mechanical seal parts, and flexible piping connectors. Several pumps made of these materials have reached the market in recent years. Problems associated have centered around these materials tendency to cold flow under pressure, and their high coefficients of expansion compared to the metallic components of the unit. Pumps may be made of heavy solid sections, as illustrated in Figure 2.3 or may use more conventional metallic components lined with the fluorocarbon materials as shown in Figure 2.4.

HEAD AND PRESSURE

To understand the performance and application of centrifugal pumps, it is essential that we grasp fully the meaning of the term "head" and its relationship to pressure.

Head

Head is a term for expressing pressure. Figure 2.5 shows the relationship between head and pressure. The gauge in the pipe near the bottom of the tank

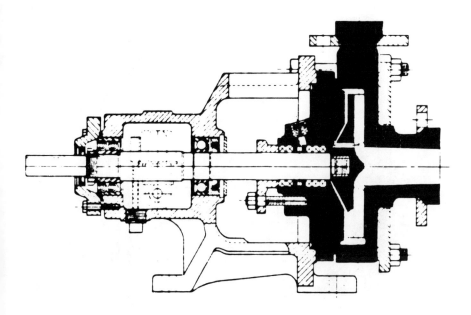

Figure 2.3 Typical construction of a centrifugal pump.

in Figure 2.5 measures the pressure created by the weight of the liquid above its centerline. We call the distance from the centerline of the gauge to the surface of the liquid as the "static head" above the gauge. The relationship between a static head and the pressure it creates is:

$$p = \frac{h \times sp.\ gr.}{2.31} \qquad \text{(Equation 2.1)}$$

or

$$h = \frac{2.31\ p}{sp.\ gr.} \qquad \text{(Equation 2.2)}$$

where:

h	= head in feet of liquid
p	= pressure, in pounds per square inch
sp. gr.	= specific gravity of the liquid

In Figure 2.5 the pressure is created by a static head of liquid. In pump

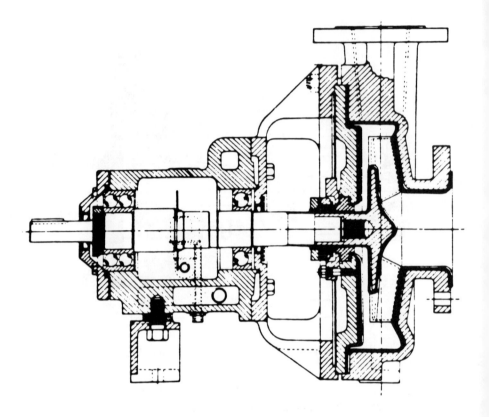

Figure 2.4. Pump with lined materials of construction.

PRESSURE & HEAD

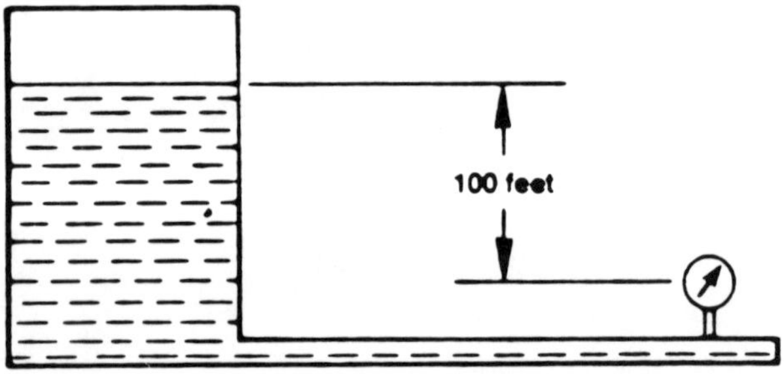

100 feet

Figure 2.5.

applications we frequently encounter pressure that is not created by static head. However, regardless of its source, any pressure can be converted into units of equivalent head using Equation 2.2.

Head & Energy

One of the advantages of using "head, in feet of liquid" to denote pressure is that it is equal to the foot-pounds of the pressure energy available from each pound of liquid. For instance, we can say that each pound of liquid at gauge level in Figure 2.5 has 100 foot-pounds of pressure energy due to the 100 foot head of liquid above it. This concept of head as an indication of energy, enables us to consider the change of pressure due to flow, as the conversion of energy.

VELOCITY HEAD AND FRICTION HEAD

Figure 2.6 shows an arrangement similar to Figure 2.5 except that the liquid is allowed to flow out of the pipe at the bottom of the tank and additional liquid is supplied to the top of the tank to keep the static head constant.

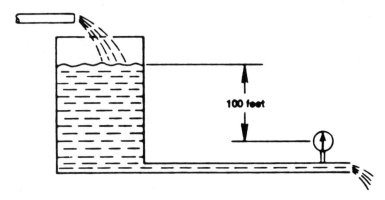

100 feet

Figure 2.6. Head measurement — incoming and outgoing liquid.

Under these conditions, the pressure gauge in the discharge line will indicate a lower head than it did in Figure 2.5. This is because, in a typical pound of liquid passing the gauge in Figure 2.6, some of the pressure energy, which was indicated as head on the same gauge in Figure 2.5 has now been converted into kinetic energy that the liquid has acquired by virtue of velocity and some of the pressure energy has been converted into heat by fluid friction in the piping ahead of the gauge.

The head, or pressure energy, that has been converted into kinetic energy is called "velocity head." It is not a "loss" of energy, but instead a change in energy form. In fact, if we cut the velocity of the liquid in half, half of the kinetic energy would convert back to pressure energy, and the head indicated by the gauge would increase accordingly.

The head that is converted into heat by fluid friction is called the "friction head" or "friction loss." From the hydraulic standpoint, it is a true loss, because heat energy is not hydraulically convertible back into pressure energy.

Going back to Figures 2.5 and 2.6, we can calculate the amount of head which was converted into kinetic energy (velocity head), if we know the velocity of the liquid passing the gauge. The relationship is as follows:

$$h_v = \frac{V^2}{2_g} \qquad \text{(Equation 2.3)}$$

where:

h_v	= velocity head, in feet of liquid
V	= velocity, in feet per second
g	= acceleration due to gravity, or 32.2 feet per second

Static Pressure

Sometimes pressure measured by a gauge, such as those in Figures 2.5 and 2.6 is called "static pressure." This is to distinguish it from "total pressure" which includes velocity head. The velocity head acts only in the direction of flow and, therefore, does not register on a gauge which is installed with its inlet perpendicular to flow.

It is important that we note the difference between the terms "static head" and "static pressure" and apply them properly. For example, in Figure 2.5 there is a "static head" (difference in elevation) of 100 feet acting at the centerline of the gauge, and the gauge measures a "static pressure" of 100 feet. On the other hand, in Figure 2.6 the "static head" is still one hundred feet but the gauge measures a "static pressure" which is less than 100 feet because some of the "static head" has been lost due to fluid friction and some has been converted to velocity head which does not register on a gauge which is installed with its inlet perpendicular to flow.

Another aspect of pressure that we should understand is that all pressures are measured with respect to some basic pressure and the resulting measurement can be identified accordingly. The pressure measurement identifications most frequently encountered in pump applications are: "absolute pressure," "gauge pressure," "vacuum," and "differential pressure."

Absolute pressure is usually expressed in "pounds per square inch, absolute," the abbreviation of which should always be "psia" not "psi." Absolute pressure measurement is relative to the complete absence of pressure, as in a perfect vacuum. In other words, absolute pressure is the amount by which the measured pressure exceeds a perfect vacuum. For example, when the atmospheric pressure is said to be "14.7 psia" (14.7 pounds per square inch, absolute), it is 14.7 pounds per square inch greater than perfect vacuum.

Gauge pressure is commonly expressed in pounds per square inch, gauge, which is usually abbreviated "psig," although sometimes the abbreviation "psi" is used. Gauge pressure is generally measured by a bourdon tube gauge or U-tube manometer with one side open to atmosphere. Gauge pressure is, therefore, relative to atmospheric pressure. To be more precise, it is the amount by which the measured pressure is greater than atmospheric pressure.

For example, when steam pressure is said to be "100 psig" (100 pounds per square inch, gauge) the steam pressure is 100 pounds per square inch greater than atmospheric pressure. If the atmospheric pressure were 14.7 psia at the gauge, then the absolute pressure of the steam would be 100 plus 14.7 or 114.7 psia.

Vacuum is a special case of gauge pressure. It, too, is relative to atmospheric pressure, but it is the amount by which the measured pressure is less than atmospheric pressure. In other words, we could say that "vacuum is negative gauge pressure." For example, a reading of "3.0 psi, vacuum" indicates that the pressure is 3.0 pounds per square inch below atmospheric pressure or −3.0 psi. gauge. So that if the atmospheric pressure at the vacuum gauge were 14.7 psia, we could calculate that 3.0 psi, vacuum is equivalent to 14.7 minus 3.0, or 11.7 psia.

The relationship between absolute, gauge, and vacuum pressures can be defined by the following verbal equations:

absolute pressure = atmospheric pressure + gauge pressure (Equation 2.4)

and

absolute pressure = atmospheric pressure − vacuum   (Equation 2.5)

and

gauge pressure = − vacuum   (Equation 2.6)

From these equations, it is obvious that each of the above terms expresses a pressure differential; however, the term "differential pressure" is usually applied to a difference in pressure between two points in a system. For example, if the pressure is 15 psig on one side of a valve and 5 psig on the other, then the differential pressure across the valve is 15 − 5, or 10 psi.

Absolute, gauge, vacuum, and differential pressure measurements can be converted into equivalent head terms, but it should be remembered that the equivalent heads are relative to the same basic pressures as were the original measurements.

CENTRIFUGAL PUMP PRINCIPLES

A centrifugal pump adds to the pressure of liquid passing through it by increasing velocity of the liquid. Figure 2.7 with views of a typical centrifugal pump, shows how this happens.

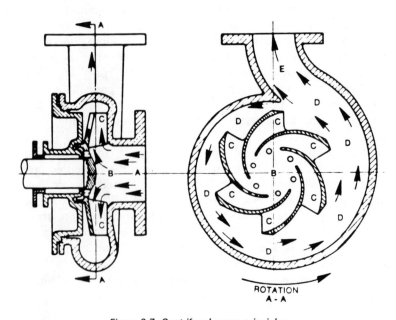

Figure 2.7. Centrifugal pump principles.

The liquid enters the pump at the suction flange, A. At this point, the liquid velocity is essentially the same as in the pipe leading to the pump. From A, the liquid flows into the impeller eye, B, where it is picked up by the impeller vanes, C. The vanes accelerate the liquid in the direction of the impeller rotation so that as it leaves the impeller, the liquid velocity approaches that of the vane tips. The casing, D, guides the liquid to the discharge neck, E, which converts some of the velocity energy to pressure energy by slowing the liquid from the casing velocity to the discharge pipe velocity.

Total Head

The sum of the gauge pressure head and the velocity head at the discharge flange minus the sum of the corresponding heads at the suction flange equals the energy (in foot-pounds) added per pound of liquid pumped and is called the "total head" developed by the pump. In equation form:

$$H = (h_{gd} + h_{vd}) - (h_{gs} + h_{vs})$$ (Equation 2.7)

where:

H   = total head
h_{gd} = discharge gauge head
h_{vd} = discharge velocity head
h_{gs} = suction gauge head
h_{vs} = suction velocity head

CENTRIFUGAL PUMP PERFORMANCE CHARACTERISTICS

The performance characteristics of a centrifugal pump are clearly defined by its performance curves. Figure 2.8 shows the performance curves of a typical centrifugal pump operating at 1750 rpm. A similar set of curves can be developed for any speed at which the pump is operated and, for most pumps a set of curves is published for each of the common induction motor speeds.

In Figure 2.8, the curve marked "H" shows the total head developed at any flow rate. For example, this pump develops a total head of 50 feet when pumping 500 gallons per minute.

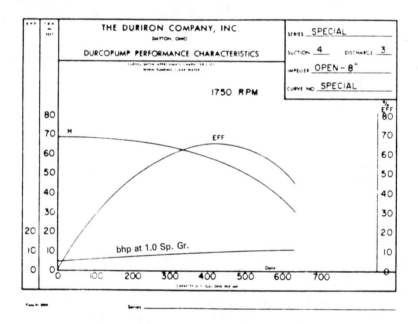

Figure 2.8. Performance curve of a centrifugal pump.

The curve marked "bhp" shows the horsepower required by the pump for any flow rate of the liquid with a specific gravity of 1.0. In the example cited above, the pump requires 10.0 horsepower to deliver 500 gpm of water. If liquid with a specific gravity of 1.5 were pumped instead of water, the horsepower required would be 1.5 times 10.0 or 15.0 horsepower.

The third curve marked "EFF," shows the pump's efficiency, in percent, at any capacity. At 500 gpm, this pump is 63% efficient.

Pump efficiency is determined mathematically. It is equal to the rate at which the pump imparts energy to the liquid, divided by the rate at which the pump requires energy, the resulting value being multiplied by 100 to put it in terms of percent. We determine the rate at which the pump imparts energy, by converting the capacity (Q) from gallons per minute into pounds per minute, and then multiplying by the total head (H) to get ft-lb per minute. The energy rate required by the pump is its bhp, which we also convert to ft-lb per minute.

Mathematically, this derivation is as follows:

$$EFF = \frac{\text{Imparted Energy Rate}}{\text{Required Energy Rate}} \times 100$$

$$= \frac{Q \times 8.33 \times sp.gr. \times H}{bhp \times 33,000 \times sp.gr.} \times 100$$

(Equation 2.8)

Notice that the specific gravity appears in both the numerator and the denominator of the expression and, therefore, cancels out, to give us the following equation for efficiency:

$$EFF = \frac{Q \times H}{bhp \text{ at } 1.0 \text{ Sp. Gr.} = 3960} \times 100 \qquad \text{(Equation 2.9)}$$

In our example, we find that:

$$EFF = \frac{500 \times 50 \times 100}{3960 \times 10} = 63\%$$

Two useful characteristics of a centrifugal pump can be observed in Figure 2.8. First, that the head produced by the pump is limited to that which is developed at shutoff (zero gpm). Therefore, even if there were a complete blockage somewhere in the discharge line of the pump, the pressure on the equipment between the pump and the blockage would not exceed the shutoff head of the pump.

Second, the horsepower decreases as the flow is throttled back to shutoff. This means that, in the event of a blockage, as described above, the motor driving the pump would not be overloaded.

A third characteristic, not apparent in Figure 2.8 is that the centrifugal pump produces a steady flow. This, combined with the centrifugal pump's low starting torque, makes a very easy load for any driving unit.

TOTAL HEAD REQUIRED BY A PIPING SYSTEM

The basic purpose of any pump is to move liquid from one point to another through a system of piping. Usually this system consists of suction and discharge tanks, a pump, and interconnecting piping made up of pipe, valves, elbows and tees. The energy required to move a pound of liquid, at the desired flow rate, from the suction tank to the discharge tank is called "the total head of a system" or, more commonly, "the system head." The pump has to supply this energy, in other words, the total head developed by the pump has to equal the total head required by the system.

Normally, the system head is divided into two parts to simplify its calculation. The head tending to move liquid from the liquid level in the suction tank to the pump is called the "total suction head." The head tending to prevent flow of the liquid from the pump to the discharge tank is called the "total discharge head."

Both of these total heads can be further subdivided into three contributing heads. These are: static head, surface pressure and friction head.

Figure 2.9 depicts the static suction head and static discharge head of a simple pumping system.

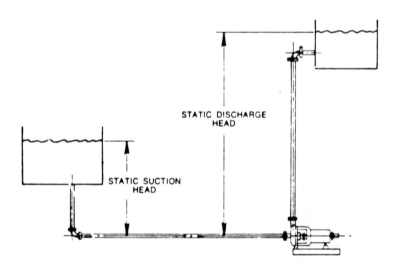

Figure 2.9. Simple Piping System

Surface Pressure

Pumping systems often begin or terminate in a tank which is under a pressure other than atmospheric. This pressure on the liquid surface affects the system head. The surface pressure, converted to feet of liquid, is called either the "suction surface pressure" or the "discharge surface pressure," dependent, of course, upon which surface it acts. Surface pressure can be expressed in "feet of liquid, gauge" or "feet of liquid, absolute" as described earlier.

Friction Head

Friction head is the energy required (per pound of liquid pumped) to overcome friction and turbulence losses that occur as the liquid flows through the piping system. Suction friction head is the sum of the energy losses occurring in the suction line. Discharge friction head is the corresponding sum of energy losses in the discharge line. The value of these losses varies with pipe length, diameter and internal finish, with the quantity of flow, and with the type of valves and fittings. The Pipe Friction Manual, published by the Hydraulic Institute, gives complete and detailed descriptions of how to calculate the friction loss in pipe, valves, and fittings.

FRICTION HEAD LOSS IN VALVES AND FITTINGS

The loss of head in a valve or fitting may be estimated by the use of this formula:

$$h_f = K \frac{V^2}{2g} \qquad \text{(Equation 2.10)}$$

where:

h_f = frictional resistance, in feet of fluid
V = average velocity, in feet per second, in a pipe of corresponding diameter
g = acceleration of gravity, 32.2 feet per second, per second
K = resistance coefficient for the valve or fitting, from the Pipe Friction Manual

One other method of calculating the loss in pipe fittings and valves is sometimes used. This method is called "the equivalent pipe length method." Table 2.1 indicates equivalent lengths of pipe fittings. Method involves adding the straight pipe lengths that are equivalent in flow resistance to pipe fittings of various types and sizes. The equivalent length of straight pipe for each fitting is added to the actual length of straight pipe and the losses are then calculated on the basis of one piece of pipe.

Table 2.1. Equivalent Length of New Straight Pipe for Valves and Fittings for Turbulent Flow Only

FLANGED FITTINGS

ITEM (C.S.)		PIPE SIZE														
		1	1¼	1½	2	2½	3	4	5	6	8	10	12	14	16	18
90° ELL		1.6	2.1	2.4	3.1	3.6	4.4	5.9	7.3	8.9	12	14	17	18	21	23
Long Radius ELL		1.6	2.0	2.3	2.7	2.9	3.4	4.2	5.0	5.7	7.0	8.0	9.0	9.4	10	11
45° ELL		.81	1.1	1.3	1.7	2.0	2.6	3.5	4.5	5.6	7.7	9.0	11	13	15	16
TEE-Line Flow		1.0	1.3	1.5	1.8	1.9	2.2	2.8	3.3	3.8	4.7	5.2	6.0	6.4	7.2	7.6
TEE-Branch Flow		3.3	4.4	5.2	6.6	7.5	9.4	12	15	18	24	30	34	37	43	47
180° Return Bend	Reg.	1.6	2.1	2.4	3.1	3.6	4.4	5.9	7.3	8.9	12	14	17	18	21	23
	Long Rad.	1.6	2.0	2.3	2.7	2.9	3.4	4.2	5.0	5.7	7.0	8.0	9.0	9.4	10	11
Globe Valve		45	54	59	70	77	94	120	150	190	260	310	390	–	–	–
Gate Valve		–	–	–	2.6	2.7	2.8	2.9	3.1	3.2	3.2	3.2	3.2	3.2	3.2	3.2
Angle Valve		17	18	18	21	22	28	38	50	63	90	120	140	160	190	210
Swing Check Valve		7.2	10	12	17	21	27	38	50	63	90	120	140	–	–	–
Bell Mouth Inlet		.18	.26	.31	.43	.52	.67	.95	1.3	1.6	2.3	2.9	3.5	4.0	4.7	5.3
Square Mouth Inlet		1.8	2.6	3.1	4.3	5.2	6.7	9.5	13	16	23	29	35	40	47	53
Re-Entrant Pipe		3.6	5.1	6.2	8.5	10	13	19	25	32	45	58	70	80	95	110

SCREWED FITTINGS

ITEM (C.S.)	PIPE SIZE						
	1	1¼	1½	2	2½	3	4
90° ELL	5.2	6.6	7.4	8.5	9.3	11	13
Long Radius ELL	2.7	3.2	3.4	3.6	3.6	4.0	4.6
45° ELL	1.3	1.7	2.1	2.7	3.2	4.0	5.5
TEE-Line Flow	3.2	4.6	5.6	7.7	9.3	12	17
TEE-Branch Flow	6.6	8.7	9.9	12	13	17	21
180° Return Bend	5.2	6.6	7.4	8.5	9.3	11	13
Globe Valve	29	37	42	54	62	79	110
Gate Valve	.84	1.1	1.2	1.5	1.7	1.9	2.5
Angle Valve	17	18	18	18	18	18	18
Swing Check Valve	11	13	15	19	22	27	38
Union or Coupling	.29	.36	.39	.45	.47	.53	.65

Formula for Sudden Enlargement:

$$h = \frac{(V_1 - V_2)^2}{2g} \quad \text{Feet of Liquid}$$

At the present time, this method is the only one available for use in determining the friction loss in valves and fittings when handling a viscous liquid. After the equivalent length of straight pipe has been selected, the friction loss can be determined by use of the charts for viscous liquids, Figures 5 through 17 of the Pipe Friction Manual.

Losses incurred in passing through other auxiliary equipment such as heat exchangers, filters, strainers, etc. should be obtained from the equipment manufacturer.

Total Discharge Head & Total Suction Head

The sources of head discussed earlier contribute to both the total discharge and total suction heads. These total heads can be calculated using the following equations:

$$h_s = h_{ss} + h_{ps} - h_{fs} \qquad \text{(Equation 2.11)}$$

where:

h_s = total suction head
h_{ss} = suction static head
h_{ps} = suction surface pressure
h_{fs} = suction friction head

and

$$h_d = h_{sd} + h_{pd} + h_{fd} \qquad \text{(Equation 2.12)}$$

where:

h_d = total discharge head
h_{sd} = static discharge head
h_{pd} = discharge surface pressure
h_{fd} = discharge friction head

Obviously, the three contributing heads must have been converted previously to feet of liquid before Equation 2.11 and Equation 2.12 can be solved.

It should also be noted that the total head calculated by either of the above equations will be in "feet of liquid, gauge" or "feet of liquid, absolute" depending on which term was used to express the surface pressure.

System Head

As stated earlier, the system head (or total head of the system) is divided into the total discharge head and the total suctionhead to simplify the calculation. After determining these contributing heads, as shown earlier, the system is evaluated using the following equation:

System head = total discharge head — total suction head

$$H = h_d - h_s \qquad \text{(Equation 2.13)}$$

It should be noted that the system head is a differential value; thus to solve equation 4.4 correctly, both h_d and h_s must be expressed in the same units. In other words, both must be "gauge" or both must be "absolute" values.

It should also be understood that, because the friction components of the system head vary with flow rate, the total head calculated by Equation 2.13 occurs at the specified rate of flow for which the friction heads were determined.

Calculation Examples

Two examples have been selected to illustrate the methods for calculating total heads of typical pumping systems. The friction loss values are obtained from the Pipe Friction Manual of the Hydraulic Institute.

Example 1

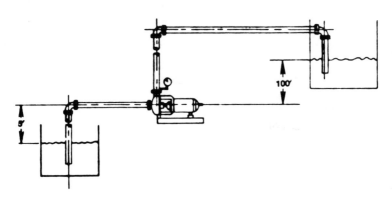

Figure 2.10.

In the first example we must determine the total head for the above system at rated flow. The pump is to transfer 100 gallons per minute of heavy acid, 1.7 specific gravity, water viscosity, from the suction tank to the discharge tank.

The liquid level in the suction tank is 5 feet below the centerline of the pump. The friction losses in the suction line are 4 feet at the rated flow. The liquid level in the discharge tank is 100 feet above the suction nozzle centerline. Friction losses in the discharge line are 25 feet at the rated flow.

Solution

To solve the problem you divide the system into two sections with the pump as the dividing line.

a. Total suction head calculation

1. The static suction head is negative because the liquid level in the suction tank is below the centerline of the pump. Thus:

$$h_{ss} = -5 \text{ Feet}$$

2. The suction tank is open, therefore, the suction surface pressure equals atmospheric pressure, or:

$$h_{ps} = 0 \text{ feet, gauge}$$

3. The suction friction head is given as:

$$h_{fs} = 4 \text{ feet at rated flow}$$

4. The total suction head using equation 2.11

$$h_s = h_{ss} + h_{ps} - h_{fs} = -5 + 0 - 4$$
$$= -9 \text{ feet of liquid, gauge, at rated flow}$$

Note that the total suction head is a gauge value because the suction surface pressure is a gauge value.

b. Total discharge head calculation

1. The static discharge head is:

$$h_{sd} = 100 \text{ feet}$$

2. The discharge tank is also open to atmospheric pressure thus:

$$h_{pd} = 0 \text{ feet, gauge}$$

3. The discharge friction head is given as:

$$h_{fd} = 25 \text{ feet at rated flow}$$

4. The total discharge head per Equation 2.12 is

$$h_d = h_{sd} + h_{pd} + h_{fd} = 100 + 0 + 25$$
$$= 125 \text{ feet of liquid, gauge, at rated flow}$$

c. Total system head calculation
   From Equation 2.13 we find:

$$H = h_d - h_s = 125 - (-9) = 134 \text{ feet of liquid at rated flow}$$

Note that both the total suction head, h_s and the total discharge head, h_d must be in the same units, in this case "feet of liquid, gauge," to solve Equation 2.13 correctly.

Example 2

In this next case we are to determine the total head and plot the curve for the following system. This pump is used to unload tank trucks of 1.2 specific gravity, water viscosity, liquid. The liquid will be pumped into a process line that is at 10 psi, pressure. The desired capacity is 160 gpm. Priming will be accomplished by an auxiliary means.

The suction line is made up of 2 swivel joints, 5—90° flanged elbows, 1 flanged gate valve, a bell mouth inlet, and 15 feet of pipe. All of this piping is 3 inch, Schedule 40, new steel pipe and all of the fittings are 3 inch. The minimum level in the tank truck is 5 feet below the pump centerline.

The discharge piping is made up of 2 flanged gate valves, 3—90° screwed elbows, 1 flanged tee (flow is line to line) and 50 feet of pipe. This piping is 2 inch, Schedule 40 steel pipe and all of the fittings are 2 inch. The process line is located 10 feet above the pump centerline.

Solution

Divide the system into two parts with the pump as the dividing line.
   a. Total suction head calculation
      1. Pump must be capable of operating down to the minimum level in the tank truck under which condition the static suction head is:

$$h_{ss} = -5 \text{ feet}$$

      2. The truck tank is open to atmospheric pressure during unloading, therefore, the suction surface pressure is:

$$H_{ps} = 0 \text{ feet, gauge}$$

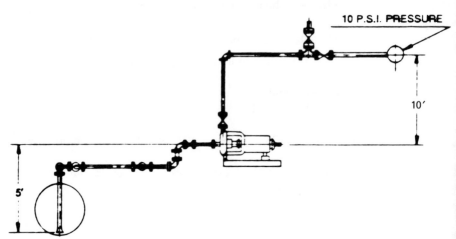

Figure 2.11. Example 2.

3. Since friction loss varies with the flow rate, we will determine the suction friction loss for each of several arbitrarily selected capacities, using Table 2.1 to obtain equivalent lengths. Two two swivel joints will be treated as four long radius flanged elbows.

Equivalent length of new 3″ pipe for:

1 — Bell mouth inlet	= 0.7 feet
5 — 3″ flanged elbows	= 22.0 feet
1 — 3″ flanged gate valve	= 2.8 feet
2 — 3″ swivel joints	= 13.6 feet
Total equivalent length	39.1 feet
Total equivalent length	= 39.1 feet
Actual length of 3″ pipe	= 15.0 feet
Total length of 3″ pipe (equivalent plus actual)	= 54.1 feet

The friction loss at each capacity is equal to the friction loss per 100 feet multiplied by the total length of pipe divided by 100.

Capacity gpm	Friction Factor (feet/100 feet)	Total Pipe Length (feet)	Suction Friction Loss h_{fs} (feet)
40	0.44	54.1	0.2
80	1.57	54.1	0.8
120	3.37	54.1	1.8
160	5.81	54.1	3.1
200	8.90	54.1	4.8

4. The total suction head is calculated by solving Equation 2.11 for each of the flow rates selected above:

$$h_s = h_{ss} + h_{ps} - h_{fs} \qquad \text{(Equation 2.11)}$$

Capacity (gpm)	Static Suction Head h_{ss} (feet)	Suction Surface Pressure h_{ps} (feet, gauge)	Suction Friction Loss h_{fs} (feet)	Total Suction Head h_s (feet, gauge)
40	−5	0	0.2	−5.2
80	−5	0	0.8	−5.8
120	−5	0	1.8	−6.8
160	−5	0	3.1	−8.1
200	−5	0	4.8	−9.8

b. The total discharge head calculation
   1. The static discharge head is:

$$h_{sd} = 10 \text{ feet}$$

   2. The discharge surface pressure is converted to feet by using Equation 2.2 as follows:

$$h_{pd} = \frac{p \times 2.31}{\text{sp. gr.}} = \frac{10 \times 2.31}{1.2} = 19.3 \text{ feet, gauge}$$

   3. The discharge friction loss will also be calculated for each of the selected capacities in the same manner as the suction friction loss calculation.

Equivalent length of new 2″ pipe for:

2 — 2″ flanged gate valves = 2 × 2.6	=	5.2 feet
3 — 2″ screwed elbows = 3 × 8.5	=	25.5 feet
1 — 2″ flanged tee	=	1.8 feet
Total equivalent length of 2″ pipe	=	32.5 feet
Actual length of 2″ pipe	=	50.0 feet
Total length of 2″ pipe (equivalent plus actual)	=	82.5 feet

In addition to the above, there is a friction loss at the sudden enlargement where the discharge pipe enters the process line. From Table 32 (b) of the Pipe Friction Manual, we see that this loss equals the velocity head which is found in Table 9 of the Pipe Friction Manual.

Capacity gpm	Total Pipe Length (feet)	Pipe Friction Factor (ft./100 ft.)	Friction Loss In Pipe (feet)	Friction Loss At Enlarge- ment (feet)	Discharge Friction Head h_{fs} (feet)
40	82.5	3.1	2.6	0.2	2.8
80	82.5	11.4	9.4	0.9	10.3
120	82.5	24.7	20.4	2.1	22.5
160	82.5	43.0	35.5	3.6	39.1
200	82.5	66.3	54.7	5.7	60.4

4. The total discharge head is calculated by solving Equation 2.12 for each of the above flow rates.

$$h_d = h_{sd} + h_{pd} + h_{fd} \qquad \text{(Equation 2.12)}$$

Capacity (gpm)	Static Discharge Head h_{sd} (feet, gauge)	Suction Surface Pressure h_{pd} (feet, gauge)	Discharge Friction Loss h_{fd} (feet)	Total Discharge Head h_d (feet, gauge)
40	10	19.3	2.8	32.1
80	10	19.3	10.3	39.6
120	10	19.3	22.4	51.7
160	10	19.3	39.1	68.4
200	10	19.3	60.4	89.7

c. Total system head calculation
  1. The total system head is determined for each of the selected capacities using Equation 2.13

$$H = h_d - h_s$$

Capacity (gpm)	Total Discharge Head h_d (feet, gauge)	Total Suction Head h_s (feet, gauge)	Total System Head H (feet)
40	32.1	−5.2	37.3
80	39.6	−5.8	45.4
120	51.7	−6.8	58.5
160	68.4	−8.1	76.5
200	89.7	−9.8	99.5

The calculated total system head, H, is plotted against capacity in Figure 2.12.

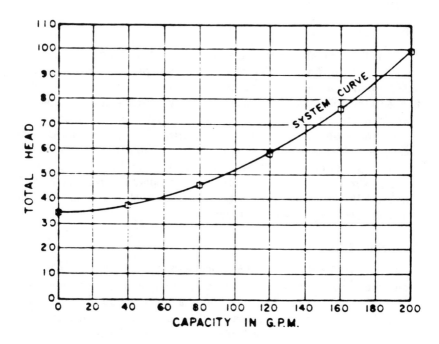

Figure 2.12. Example 2. Capacity & head relationship.

NET POSITIVE SUCTION HEAD

The one hydraulic characteristic that is least understood by the people who maintain, or operate centrifugal pumps is the Net Positive Suction Head. Briefly, NPSH has to do with the availability of liquid to the pump on the suction side.

NPSH and Cavitation

We must check the NPSH conditions for each pump application to determine whether the liquid to be pumped will vaporize inside the pump. Vaporization within a pump is called "cavitation." Cavitation reduces a pump's performance and may damage the pump. When the pump is starved for liquid, it becomes quite noisy and the wet end parts will become badly eroded and worn.

To understand the occurrence of cavitation, it is important to remember that a liquid will vaporize at a comparatively low temperature if its pressure is reduced sufficiently. Water, for instance, will vaporize at 100°F. if it is exposed to a vacuum of 28 inches of mercury. The pressure at which a liquid will vaporize is called its "vapor pressure."

NPSH Conditions

From the preceding paragraph, we see that a reduction in pressure can cause a liquid to vaporize if it is close to its vapor pressure. The pressure on the liquid entering a centrifugal pump is reduced as it moves from the suction flange to the point at which it receives energy from the impeller. Obviously, we must compare this reduced pressure to the vapor pressure entering the pump to determine whether the liquid will vaporize. This is what we do when we check the NPSH conditions of an application. We call the proximity of the liquid to its vapor pressure its "available NPSH" and the pressure reduction inside the pump, the "required NPSH." We compare the available NPSH to the required NPSH. When the available NPSH is equal to or greater than the required NPSH the pump will not cavitate.

Available NPSH

A more precise definition of available NPSH is "the difference between the total suction head the the vapor pressure of the liquid, in feet of liquid, at the suction flange." We can measure the total suction head of the pump and we can find vapor pressure from the liquid temperature. The difference between these two values is the available NPSH. The following equation is the mathematical expression of the definition for available NPSH:

$$H_{sv} = h_{sa} - h_{vpa} \qquad \text{(Equation 2.14)}$$

where:

h_{sv} = available net positive suction head, in feet of liquid
h_{sa} = total suction head, in feet of liquid, absolute
h_{vpa} = vapor pressure of liquid at suction nozzle, in feet of liquid, absolute

Required NPSH

The required NPSH can be defined as "the reduction in total head as the liquid enters the pump." Figure 2.13 illustrates how this reduction in head takes place.

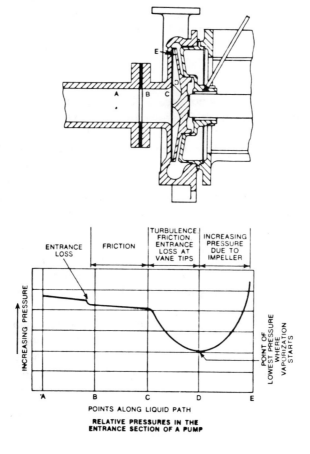

Figure 2.13.

POINTS ALONG LIQUID PATH

RELATIVE PRESSURES IN THE ENTRANCE SECTION OF A PUMP

In the pump shown in Figure 2.13, there may be an entrance loss at the suction flange, A. The suction nozzle, B, even though very short, has a loss due to friction. The turn that the liquid makes at C in going from axial flow in the suction nozzle to outward radial flow in the impeller is accompanied by both friction loss and turbulence. The leading edges of the vanes act as obstructions in the path of the liquid and an entrance loss occurs at each vane edge. Any pre-rotation of the liquid as it enters the impeller changes the inlet angle and results in more turbulence. All of these losses occur before the liquid is acted upon positively by the impeller vanes. Once the liquid is in the impeller, with the vanes pushing from behind, the pressure starts to increase and eventually reaches the full discharge head. A theoretical plot of liquid pressure in going through this section of the pump is shown in Figure 2.13.

The pump manufacturer determines the required NPSH for each pump by test procedures, and plots the results on the standard performance curves for that pump.

NPSH Problems

If the available NPSH is not greater than that required by the pump, many serious problems can result. There will be a marked reduction in head and capacity, or even a complete failure to operate. Excessive vibration can occur when sections of the impeller are handling vapor and the other sections handling liquid. Probably the most serious problem is pitting and erosion of the pump parts, resulting in reduced life. This is caused by the collapse of vapor bubbles as they pass to the regions of higher pressure. This cavitation phenomenon is usually accompanied by excessive noise and vibration. As the vapor bubbles collapse, the adjacent walls are subjected to a tremendous shock from the inrush of liquid into the cavity left by the bubble. This shock actually flakes off small bits of metal and the parts take on the appearance of having been badly eroded. This erosion shows up not at the point of lowest pressure where the bubble is formed, but further downstream where the bubble collapses.

The energy expended in accelerating the liquid to high velocity in filling the void left by the bubble is a loss, and causes the drop in head associated with cavitation. The loss in capacity is the result of pumping a mixture of vapor and liquid instead of liquid. Water, for example, at 70 °F. increases in volume about 54,000 times when vaporized, and thus even a slight amount of cavitation will reduce the capacity.

A pump operating with insufficient available NPSH will often pump spurts of liquid. This is caused by the following chain of events. As the pump is started, the liquid accelerates in the suction nozzle until it reaches the capacity at which it is to operate. As it accelerates, the friction losses increase and lower the absolute pressure until the liquid flashes into vapor. As soon as this happens, the pumping action is reduced, and the flow decreases. With the decreased flow, the losses are lower, the absolute pressure is higher, the liquid does not vaporize, and the pump starts to pump again. This increases the flow, reduces the pressure, etc., until the whole cycle is repeated. This results in an erratic flow rate with spurts of liquid being thrown from the discharge pipe.

Calculating Available NPSH of a Piping System

There are five typical pump installations for which the available NPSH should always be calculated. These are: 1) when the pump is installed an appreciable height above the liquid level; 2) when the pump takes suction from a tank under vacuum; 3) when the liquid has a high vapor pressure; 4) when the suction line is unusually long; and 5) when the pumping system is at an altitude considerably above sea level (where the atmospheric pressure is reduced). The available NPSH can be calculated by use of the following formula:

$$h_{sv} = h_{psa} + h_{ss} - h_{fs} - h_{vpa} \qquad \text{(Equation 2.15)}$$

where:

h_{sv} = available net positive suction head in feed of liquid.

h_{psa} = suction surface pressure, in feet of liquid, absolute, on the surface of the liquid from which the pump takes its suction. This will be the atmospheric pressure, in the case of an open tank, or the absolute pressure above the liquid in a closed tank.

h_{ss} = static suction head, in feet of liquid. In other words, the height, in feet, of the liquid surface in the suction tank above or below the pump centerline. (Positive if the liquid level is above the pump, negative if the liquid level is below the pump).

h_{fs} = friction head loss, in feet of liquid, between the liquid surface in the suction tank and the suction flange of the pump.

h_{vpa} = vapor pressure of the liquid, at the pumping temperature, in feet of liquid, absolute.

Note that the first three terms in Equation 2.15 equal the total suction head, h_{sa}, and if we replace the first three terms with h_{sa}, we get Equation 2.14 which is a mathematical definition of available NPSH.

Each calculation of available NPSH for a piping system requires the following five steps:

Step 1: Determine the suction surface pressure, h_{psa}.

This is the pressure on the surface of the liquid in the suction tank. When the suction tank is open, the suction surface pressure equals atmospheric pressure. When the suction tank is closed the pressure on the surface of the liquid must be measured.

Step 2: Determine the static suction head, h_{ss}.

This is the height, in feet, of the liquid surface in the suction tank above or below the pump centerline. When the liquid level is below the pump centerline, the static suction head is a negative value.

Step 3: Determine the suction friction head, h_{fs}.

This is the sum of all the friction losses in the suction line from its inlet to the suction flange of the pump, at the specified flow rate. The friction loss factors are from the Pipe Friction Manual of the Hydraulic Institute. [Pipe friction per 100 feet of pipe and velocity head are found in Tables 1 through 31 in that Manual and K factors for pipe fittings are found in Tables 32 (a) and 32 (b)].

Step 4: Determine the vapor pressure, h_{vpa}, of the liquid at the pumping temperature, in feet of liquid, absolute.

Step 5: Calculate the available NPSH from Equation 2.15 using the values determined in steps 1 through 4.

In the following examples, the abbreviation "PFM" is used for Pipe Friction Manual.

Example 3

Suction lift.

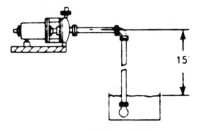

Figure 2.14. Example 3.

The pump shown in Figure 2.14 is handling 100 gpm of water at 60°F. The suction lift is 15 feet. The suction line consists of 25 feet of 2 inch steel pipe with one 90° flanged elbow and one 2 inch foot valve. The barometric (atmospheric) pressure is 29.96 inches of mercury.

The available NPSH at 100 gpm is found as follows:

Step 1: $h = \dfrac{\text{inches of mercury} \times 1.13}{\text{sp. gr.}}$ $\qquad h_{psa} = \dfrac{29.96 \times 1.13}{1.0}$

$$= 33.9 \text{ feet, absolute}$$

Step 2: $h_{ss} = -15.0$ feet

Step 3: Head loss per 100 feet = 17.4 from Table 2-PFM

Velocity head = 1.42 from Table 2-PFM

h_f (for the pipe)	$= 25 \times 17.4/100$	$= 4.4$ feet
h_f (for the valve)	$= 0.80 \times 1.42$	$= 1.1$ feet
h_f (for the elbow)	$= 0.37 \times 1.42$	$= 0.5$ feet
	At 100 gpm, h_{fs}	$= 6.0$ feet

Step 4: h_{vpa} = 0.6 feet from steam tables.
Step 5:

$$h_{sv} = h_{psa} + h_{ss} - h_{fs} - h_{vap} = 33.9 + (-15.0) - 6.0 - 0.6$$
$$= 12.3 \text{ feet of liquid at 100 gpm}$$

If this pump were relocated at an elevation of 5,000 feet, the only term, in Equation 2.15 that would change is the suction surface pressure h_{psa}. The standard atmospheric pressure at 5,000 feet 12.2 psia

Table 2.2. Altitude vs. Atmospheric Pressure

Altitude Above Sea Level Feet	Atmospheric Pressure psia
0	14.7
1000	14.2
2000	13.7
3000	13.2
4000	12.7
5000	12.2
6000	11.7
7000	11.3
8000	10.9
9000	10.5
10000	10.1

Step 1: from equation 2.2

$$h = \frac{2.31 \text{ p}}{\text{sp. gr.,}} \qquad h_{psa} = \frac{2.31 \times 12.2}{1.0} = 28.2 \text{ feet, absolute}$$

Step 2: (As above)
Step 3: (As above)
Step 4: (As above)
Step 5: h_{sv} = 28.2 + (-15.0) - 6.0 - 0.6 = 6.6 feet at 100 gpm

We frequently need to determine the available NPSH at the suction of a pump after it is installed. If we know the flow rate and the vapor pressure of the liquid and can measure the suction gauge pressure, we can calculate the available NPSH using the following equation.

$$H_{sv} = h_{gs} + h_a + h_{vs} - h_{vpa} \qquad \text{(Equation 2.16)}$$

where:

h_{sv} = available net positive suction head in feet of liquid
h_{gs} = suction gauge pressure, in feet of liquid, gauge
h_a = atmospheric pressure, in feet of liquid, absolute
h_{vs} = suction velocity head, in feet of liquid
h_{vpa} = vapor pressure of the liquid, in feet of liquid, absolute

Note that the first three terms in Equation 2.16 equal the total suction head, h_{sa}, and we can replace the first three terms with h_{sa}.

Calculation of available NPSH using the above equation involves the following five steps:

Step 1: Convert the suction gauge pressure, h_{gs}, to feet of liquid, gauge using Equation 2.2.

Step 2: Convert the atmospheric pressure, h_a, to feet of liquid, absolute.

Step 3: Find the suction velocity head, h_{sv}, from Tables 1 through 31-PFM.

Step 4: Convert the vapor pressure, h_{vpa} to feet of liquid, absolute.

Step 5: Solve Equation 15 using the values determined in Steps 1 through 4.

Example 4

The following example illustrates the above calculation. Determining available NPSH by gauge reading.

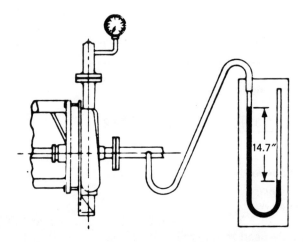

Figure 2.15. Example 4: Available NPSH by gauge reading.

This pump is handling 100 gpm of acid with a specific gravity of 1.7 and a vapor pressure of 0.20 psia at pumping temperature. A mercury manometer connected to the 2 inch pipe at the suction nozzle reads 14.7 inches, vacuum and the barometric pressure is 736 millimeters of mercury.

Find the available NPSH as follows:

Step 1:

$$h = \frac{\text{inches of mercury} \times 1.13}{\text{sp. gr.}}$$

$$h_{gs} = \frac{-14.7 \times 1.13}{1.7} = 9.8 \text{ feet, gauge}$$

Step 2:

$$h = \frac{\text{millimeters of mercury}}{22.4 \times \text{sp. gr.}}$$

$$h_{a} = \frac{736}{22.4 \times 1.7} = 19.3 \text{ feet absolute}$$

Step 3: $h_{vs} = 1.4$ feet at 100 gpm from Table 9-PFM

Step 4: from equation 2.2

$$h = \frac{2.31p}{\text{sp. gr.}}$$

$$h_{vpa} = \frac{2.31 \times 0.20}{1.7} = 0.3 \text{ feet absolute}$$

Step 5: from equation 2.16

$$h_{sv} = h_{gs} + h_{a} + h_{vs} - h_{vpa} = -9.8 + 19.3 + 1.4 - 0.3$$

$$= 10.6 \text{ feet at 100 gpm}$$

Testing for Required NPSH

To test a pump for its NPSH requirement, we gradually reduce the total suction head until the liquid just starts to vaporize in the impeller, causing a drop in the total head developed by the pump. A three percent drop in total head is generally considered to indicate incipient cavitation. At this point, we record the suction gauge pressure, the flow rate, the pumping temperature, the barometric pressure and the pump rpm. This information enables us to solve Equation 2.16 to determine the available NPSH.

When a pump is at the point of incipient cavitation, the NPSH available at its suction flange has been reduced to the degree that it just equals the NPSH required by the pump. In other words, solving Equation 15 for this condition gives us both the NPSH available and the NPSH required.

The example 5 which follows illustrates the determination of one point on a typical NPSH curve. A series of these points must be plotted against capacity in order to describe the NPSH required by the pump at all capacities.

Example 5

Determine the NPSH of 2 × 1½ pump when handling 200 gpm of clear water at 90°F. The barometer reading at the test site is 29.0 inches of mercury. The lowest suction gauge pressure that could be obtained when the total head developed by the pump was within 3 percent of its normal total head, was 17.7 inches of mercury, vacuum.

Step 1: $h = \dfrac{\text{inches of mercury} \times 1.13}{\text{sp. gr.}}$

$$h_{gs} = \frac{-17.7 \times 1.13}{0.996} = -20.1 \text{ feet, gauge}$$

Step 2: $h = \dfrac{\text{inches of mercury} \times 1.13}{0.996}$

$$h_a = \frac{29.0 \times 1.13}{0.996} = 32.9 \text{ feet, absolute}$$

Step 3: $h_{vs} = 5.7$ feet at 200 gpm from Table 9-PFM
Step 4: $h_{vpa} = 1.6$ feet, absolute
Step 5:

$$h_{sv} = h_{gs} + h_a + h_{vs} - h_{vpa} = -20.1 + 32.9 + 5.7 - 1.6 = 16.9 \text{ feet}$$

This value is then plotted as the required NPSH of the pump at the 200 gpm capacity point.

VISCOSITY

Characteristics of Viscous Liquids

The viscosity of a liquid is defined as a measure of its resistance to flow. It may be considered as the internal friction resulting when one layer of a fluid is made to move in relationship to another layer. The so called "thick" liquids, such as automobile transmission lubricants, molasses, varnish, etc. have high viscosities. "Thin" liquids, such as water, gasoline and hydrochloric acid, have low viscosities. The performance of a centrifugal pump is adversely affected when the liquid being handled has a viscosity higher than that of water. Since flow is associated with shearing forces between sections of a liquid, more energy is required to produce flow in a viscous liquid than in a non-viscous liquid. This added energy is evidenced by an increase in horsepower and a reduction in head, capacity, and efficiency of the pumping unit.

Liquids that have high specific gravities are not necessarily highly viscous. Moreover, the viscosity of a liquid varies appreciably with changes in temperature, but very little with changes in pressure. A common example of this is maple syrup. When chilled, it is very stiff and hard to pour. A slight warming thins it, however, allowing it to flow quite freely. For this reason, the viscosity number must always be qualified by giving the temperature at which this viscosity was determined. In selecting a centrifugal pump, it is imperative that the viscosity value used be in agreement with the pumping temperature of the liquid. Inaccuracies in this respect could result in the selection of the wrong size pump and motor.

Measuring Viscosity

The viscosity of a liquid is determined by an instrument called a viscosimeter. Liquids of low to medium viscosity are measured by the Saybolt Universal Viscosimeter, and liquids of high viscosity are measured by the Saybolt Furol Viscosimeter. A measured volume of liquid is allowed to flow through an orifice of specified proportions and the time of efflux noted. This time in seconds, is then called the SSU number (Seconds Saybolt Universal) or SSF numbers (Seconds Saybolt Furol). These numbers have been so widely accepted that they are often used in place of, or in addition to, the actual viscosity in centistokes. The basic unit for measuring absolute or kinetic viscosity. Most of the charts published by the Hydraulic Institute show viscosities in both scales.

Tables 36 (a) through (f) of the Pipe Friction Manual list the viscosities of many common liquids at several temperatures. Upon examination you will notice the wide variation in viscosity possible with comparatively small variation in temperature.

The friction loss in pipe, valves, and fittings increases with viscosity.

The hydraulic losses that occur in a pump are due to the viscosity of the liquid being handled. The standard performance curve is valid only when the pump is handling liquid with a viscosity close to or less than 1.0 centistoke.

When handling a liquid with an appreciably higher viscosity, the pump will not perform as shown on the standard curve. The head, capacity, and efficiency will be less, while the horsepower will increase.

Unfortunately, there is no acceptable analytic method of establishing the pump performance when the liquid has a viscosity other than that of water. A large number of experiments have been conducted as a means of developing an empirical method of predicting this performance. The data from these tests have been formulated into charts or nomographs so that the pump performance can be estimated for liquids of practically any normal viscosity.

Viscosity Correction Nomograph

Figure 2.16 provides a means of estimating the viscous performance of a centrifugal pump when its water performance is known. If exact viscous performance data are required, performance tests should be made using the viscous liquid in question.

This chart is to be used only for standard centrifugal pumps, not for mixed or axial flow pumps. It is limited to Newtonian liquids, i.e., fluids that are unaffected by the magnitude and kinds of motion to which they are subjected. Slurries, paper stock, dilatant or thixotropic liquids may produce widely varying results. In addition, the chart is only to be used where there is adequate NPSH available because the exact effect of viscosity on NPSH required is not predictable.

The following symbols and equations are employed in the application of Figure 2.16 to pump viscosity problems.

$$Q_{vis} = C_Q \times Q_w \qquad \text{(Equation 2.17)}$$
$$H_{vis} = (1 - C_H \times Q_w [Q_w]) \times H_w \qquad \text{(Equation 2.18)}$$
$$BHP_{vis} = C_{HP} \times \text{sp. gr.} \times BHP_w \qquad \text{(Equation 2.19)}$$

where

Q_{vis}	= viscous capacity, in gpm
H_{vis}	= viscous total head, in feet
BHP_{vis}	= viscous brake horsepower
Q_w	= water (non-viscous) capacity, in gpm
$[Q_w]$	= water capacity at best efficiency, in gpm
H_w	= water total head, in feet
BHP_w	= water brake horsepower
C_Q	= viscosity correction factor for capacity
C_H	= viscosity correction factor for total head
C_{HP}	= viscosity correction factor for horsepower

Correction of Water Performance Curves to Viscous Performance Curve

The viscosity correction factors obtained from Figure 2.16 enables us to derive the viscous performance curves of a pump from its standard water curves. The procedure is as follows:

Step 1. Make up a worksheet as shown in Figure 2.17 wherein, each of the vertical volumns represents the performance parameters at a given flow rate.

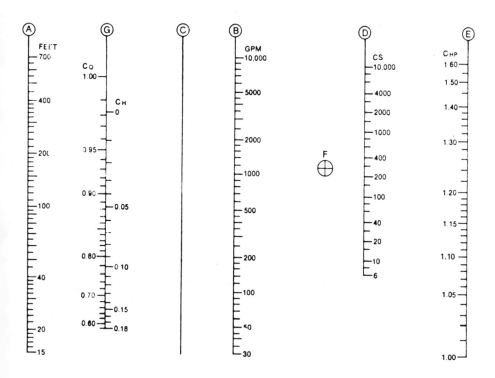

Figure 2.16. Viscosity correction nomograph.

Step 2. At the best efficiency on the water performance curves, read the capacity $[Q_w]$, the total head, H_w, and brake horsepower, BHP_w. Record their values in the appropriate spaces in colume 3 of the worksheet.

Step 3. On Figure 2.16, align a straight edge with the value of H_w, on line A and the value of $[Q_w]$ on line B. Mark line C where it intersects the straight edge. Now align the straight edge with the mark on line C and the value of the viscosity on line D. Mark line E where it intersects the straight edge. Read and record in column 3 of the worksheet, (Figure 2.17) the Horsepower Correction Factor, C_{HP}. Now align the straight edge with the mark on line E and point F. Mark line G where it intersects the straight edge. Read and record in column 3 of the worksheet, the correction factors C_Q, and C_H.

Step 4. From the water performance curves, select three or four more flow rates which are evenly spaced over the range of the curve to be plotted. Enter these values in the remaining Q_w spaces in the worksheet.

Step 5. From the water performance curves, read the total heads H_w, and the brake horsepower, BHP_w, that correspond to the flow rates selected in Step 4. Enter these rates in the designated spaces on the worksheet.

VICOSITY CORRECTION WORKSHEET					
		PERFORMANCE POINT			
	1	2	3	4	5
$Q_\wedge$			[ ]*		
C_Q					
Q_{vis}					
$H_\wedge$					
C_H					
$C_H \cdot \dfrac{Q_\wedge}{[Q_\wedge]}$					
H_{vis}					
BHP_w					
C_{rp}					
sp.gr.					
BHP_{vis}					

*Brackets [ ] indicate $Q_\wedge$ at best efficiency point

Figure 2.17.

Step 6. Calculate the values of $C_H \times Q_w/[Q_w]$ and record the results in the appropriate spaces on worksheet.

Step 7. Use Equations 2.16, 2.17, and 2.18 to compute the viscous capacities, Q_{vis}, heads, H_{vis}, and viscous brake horsepowers, BHP_{vis}. Enter the computed values on the worksheet.

Step 8. Plot the viscous performance curves using the computed viscous data from the worksheet.

The following example demonstrates the use of the above procedure.

Example 6

Assume a 4 × 3 − 13 pump operating at 1750 rpm with water performance as shown in Figure 2.18. Plot the viscous performance of the 12½″ diameter impeller when handling a liquid with an absolute viscosity of 1625 centipoise and a specific gravity of 1.23.

Solution

In order to use the Viscosity Correction nomograph, Figure 2.16, the

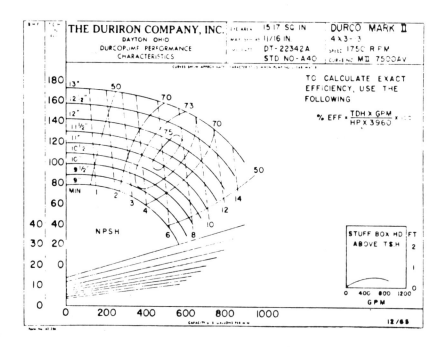

Figure 2.18.

absolute viscosity must be converted to its kinematic equivalent using Equation

$$\text{Kinematic Viscosity} = \frac{\text{Absolute Viscosity}}{\text{sp. gr.}} = \frac{1625}{1.23} = 1320 \text{ centistokes}$$

Step 1. Make up a worksheet similar to Figure 2.17.

Step 2. From the water performance curves, Figure 2.18, we find that at the maximum efficiency point of the 12½″ impeller, the capacity, $[Q_w]$, the total head, H_w, and the brake horsepower, BHP_w, are 585 gpm, 132 feet, and 26.8 hp, respectively. We enter these values in column 3 of the worksheet as shown in Figure 2.19.

Step 3. From Figure 2.16 we determine that the viscosity correction factors corresponding to $[Q_w]$ and H_w are:

$$C_H = 0.14$$

$$C_Q = 0.68$$

$$C_{HP} = 1.5$$

VICOSITY CORRECTION WORKSHEET					
	PERFORMANCE POINT				
	1	2	3	4	5
Q_w	0	400	[585]*	700	840
C_Q	0.68				
Q_{vis}	0	272	398	476	571
H_w	157	146	132	117	88
C_H	0.14				
$C_H \cdot \frac{Q_w}{[Q_w]}$	0	0.10	0.14	0.17	0.20
H_{vis}	157	132	114	97	70
BHP_w	11.8	21.9	26.8	29.7	33.2
C_{np}	1.50				
sp.gr.	1.23				
BHP_{vis}	21.8	40.4	49.4	54.8	61.3

*Brackets [ ] indicate Q_w at best
efficiency point

Figure 2.19.

We also add these values to the worksheet.

Step 4. From Figure 2.18, we pick four additional flow rates 0, 400, 700 and 840 gpm and enter them as Q_w in columns 1, 2, 4, and 5 of the worksheet.

Step 5. Again from Figure 2.18, we find the total heads (157, 146, 117, and 88 feet) and water brake horsepowers (11.8, 21.9, 29.7 and 33.2) corresponding to the selected flow rates and enter them on the worksheet as H_w and BHP_w.

Step 6. Calculate $C_H \times Q_w/[Q_w]$

At point 1 $C_H \times Q_w/[Q_w] = 0.14 \times 0/585 = 0$
At point 2 $C_H \times Q_w/[Q_w] = 0.14 \times 400/585 = 0.10$
At point 3 $C_H \times Q_w/[Q_w] = 0.14 \times 585/585 = 0.14$
At point 4 $C_H \times Q_w/[Q_w] = 0.14 \times 700/585 = 0.17$
At point 5 $C_H \times Q_w/[Q_w] = 0.14 \times 840/585 = 0.20$

Step 7. Using Equation 2.17 we calculate the viscous capacities as follows:
follows:

At point 1 $Q_{vis} = 0.69 \times 0 = 0$ gpm

At point 2 $Q_{vis} = 0.68 \times 400 = 272$ gpm
At point 3 $Q_{vis} = 0.68 \times 585 = 398$ gpm
At point 4 $Q_{vis} = 0.68 \times 700 = 476$ gpm
At point 5 $Q_{vis} = 0.68 \times 840 = 571$ gpm

Use Equation 2.18 we calculate the viscous head as follows:

At point 1 $H_{vis} = (1 - 0) \times 157 = 157$ feet
At point 2 $H_{vis} = (1 - 0.10) \times 146 = 132$ feet
At point 3 $H_{vis} = (1 - 0.14) \times 132 = 114$ feet
At point 4 $H_{vis} = (1 - 0.17) \times 117 = 97$ feet
At point 5 $H_{vis} = (1 - 0.20) \times 88 = 70$ feet

Using Equation 2.19 we calculate the viscous brake horsepower as follows:

At point 1 $BHP_{vis} = 1.5 \times 1.23 z 11.8 = 21.8$
At point 2 $BHP_{vis} = 1.5 \times 1.23 \times 21.9 = 40.4$
At point 3 $BHP_{vis} = 1.5 \times 1.23 \times 1.23 - 49.4$
At point 4 $BHP_{vis} = 1.5 \times 1.23 \times 29.7 - 54.8$
At point 5 $BHP_{vis} = 1.5 \times 1.23 \times 33.2 - 61.3$

These values are then entered on the worksheet.

Step 8. Data from the worksheet, Figure 2.19, is then plotted as shown by the dotted line curves in Figure 2.20.

Selection of a Pump for Viscous Conditions

The most common use for Figure 2.16 is in selecting a pump to satisfy viscous requirements. To do so presents an obvious problem, in that to obtain correction factors from the chart we need to know the best efficiency of the pump which is to be used. To circumvent this problem, we follow a procedure that obtains a preliminary set of viscosity factors which are admittedly incorrect but, nevertheless are sufficiently accurate to permit us to select a pump that will satisfy the viscous requirements. Once having selected the basic pump, we can obtain true factors from Figure 2.16 which allow us to select the correct impeller diameter.

The procedure employs the following equations which are transposed from Equation 2.17 and Equation 2.18:

*Brackets [] indicate Q_w at best efficiency point.

$$Q_w = \frac{Q_{vis}}{C_Q} \qquad \text{(Equation 2.17a)}$$

$$H_w = \frac{H_{vis}}{(1 - C_H \times Q_w/[Q_w])} \qquad \text{(Equation 2.18a)}$$

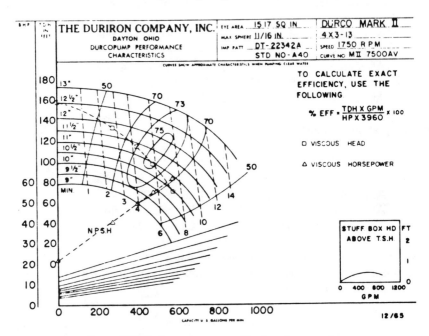

Figure 2.20. Performance curve with viscosity correction.

The procedure is as follows:

Step 1. In Figure 2.16 align H_{vis} on the A line with Q_{vis} (or 30 gpm when Q_{vis} is less than 30) on the B line. Determine the preliminary viscosity correction factors, C_{HP1}, C_{Q1} and C_{H1} corresponding to these conditions and the specified viscosity. When this procedure indicates a horsepower correction factor greater than 1.6, assume $C_{HP1} = 1.6$, $C_{Q1} = 0.58$ and $C_{H1} = 0.18$.

Step 2. Using Equation 2.17a determine the preliminary water capacity, Q_{w1}, corresponding to Q_{vis}.

Step 3. Assuming $Q_{g1}/[Q_w] = 1.0$ determine the preliminary water, head H_{w1}, corresponding to H_{vis}.

Step 4. Select a pump that satisfies Q_{w1} and H_{w1}. Read the head and capacity $H_w [Q_w]$, at the best efficiency point of the impeller curve that is nearest to, but falls below, H_{w1} at Q_{w1}.

Step 5. From Figure 2.16, obtain a secondary set of correction factors, C_{Q2} and C_{H2}, corresponding to H_w, and $[Q_w]$ determined in step 4.

Step 6. Using C_{Q2} from step 5 and Equation 2.17 compute the secondary water capacity, Q_{w2}, corresponding to Q_{vis}.

Step 7. Using $[Q_w]$ from step 4, C_{H2} from step 5 and Q_{w2} from step 6, compute a secondary water head, H_{w2}, corresponding to H_{vis}.

Step 8. Select the impeller diameter that satisfies H_{w2} and Q_{w2}. If the diameter is close to that selected in step 4, we can consider the pump selection to be complete and proceed to step 9. If the diameter differs by more than ½ inch from that selected by step 4, we must repeat steps 4 through 8, substituting the values of H_{w2} and Q_{w2} from step 8 for H_{w1} and Q_{w1} in step 4.

Step 9. Use the procedure previously described for plotting the viscous performance curves or calculating the viscous brake horsepower of the selected pump and the impeller.

The following example illustrates the above procedure.

Example 7

Select a pump to move 272 gpm of viscous liquid against a total head of 130 feet. The viscosity of the liquid is 6000 SSU and its specific gravity is 1.23. Determine the viscous brake horsepower at the operating point and at the end of the curve.

Solution

From page 74 of the Pipe Friction Manual we find that 6000 SSU = 1320 centistokes.

Step 1. From Figure 2.16 at 130 feet, 272 gpm, and 1320 centisokes we find:

$$C_{HP1} = 1.60$$
$$C_{H1} = 0.18$$
$$C_{Q1} = 0.58$$

Step 2. From Equation 2.17a

$$Q_{w1} = \frac{272}{0.58} = 469 \text{ gpm}$$

Step 3. From Equation 2.18a assuming $Q_{w1}/[Q_w] = 1.0$

$$H_{w1} = \frac{130}{(1 - 0.18 \times 1)} = 159 \text{ feet}$$

Step 4. We find that 4 × 3— 13 pump operating at 1750 rpm will satisfy Q_{w1}

and H_{w1}. From the water performance curves of this pump shown in Figure 2.18, we select the 12½″ diameter impeller as being nearest to, but falling below, H_{w1} at Q_{w1}. We read the best efficiency head and capacity of the 12½″ impeller curve as being:

$$H_w = 132 \text{ feet}$$

$$[Q_w] = 585 \text{ gpm}$$

Step 5. From Figure 2.16, we find the secondary correction factors corresponding to 132 feet and 585 gpm to be:

$$C_{HP2} = 1.5$$

$$C_{Q2} = 0.68$$

$$C_{H2} = 0.14$$

Step 6. Using Equation 2.17a

$$Q_{w2} = \frac{272}{0.68} = 400 \text{ gpm}$$

Step 7. From 2.18a using $[Q_w]$ from step 4 we find that:

$$H_{w2} = \frac{130}{(1 - 0.14 \times 400/585)} = 144 \text{ feet}$$

Step 8. Looking again at Figure 2.18, we see that the 12½″ curve satisfies Q_{w2} and H_{w2}; therefore, the 4 × 3 − 13 pump with a 12½″ impeller is the correct selection for this application.

Step 9. This step has already been completed in Example 1, since we are dealing with the same pump (4 × 3 − 13) and the viscosities are equal (6000 SSU = 1320 centistokes). From the worksheet of Example 1, we see that the viscous brake horsepower at the design point (where Q_{vis} = 272 gpm) is 40.4 and at the end of the curve (point 5) it is 61.3.

SPECIFIC SPEED AND SUCTION SPECIFIC SPEED

Specific speed is a dimensionless index number which is used to relate the hydraulic performance of a centrifugal pump to the shape and physical properties of its impeller. An indication of how impeller shape and proportions vary over the normal range of specific speeds can be seen in Figure 2.21.

SPECIFIC SPEED

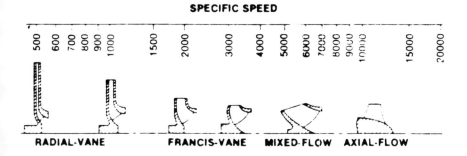

Figure 2.21.

The numerical value of specific speed is found from the following equation:

$$N_s = \frac{NQ^{1/2}}{H^{3/4}} \qquad \text{(Equation 2.19)}$$

where:

N_s = specific speed
N = rotative speed, in revolutions per minute
Q = capacity, at best efficiency, in gallons per minute
H = total head, at best efficiency, in feet

It can be shown by applying the affinity laws to the terms of the above equation that the specific speed of a given impeller is constant regardless of its rotative speed.

Suction specific speed is a dimensionless rating number which indicates the relative ability of centrifugal pumps to operate under conditions of low available net positive suction head.

The equation for suction specific speed:

$$S = \frac{NQ^{1/2}}{h_{sv}^{3/4}} \qquad \text{(Equation 2.20)}$$

where:

S = suction specific speed
N = rotative speed, in revolutions per minute
Q = capacity, at best efficiency, in gallons per minute
H_{sv} = net positive suction head, required at best efficiency, in feet

Depending on impeller design, suction specific speeds will vary in numerical value from below 4,000 to above 11,000 with the higher values indicating lower net positive suction head requirements.

Impeller Loading

In addition to the obvious torsional reaction, centrifugal pump impellers are also subjected to appreciable radial and axial loading.

Radial Loading

At the flow rate of best efficiency, in a well designed volute pump, the pressure developed by the impeller is more or less uniform about its entire circumference. The resultant radial force on the impeller is essentially zero. However, at all other flow rates, the pressure is not uniform and this pressure variation, acting on the projected impeller area, produces a resultant radial force on the impeller. At low flow rates, the pressure distribution in a volute type casing is such that the impeller surfaces closest to the discharge are acted upon by high pressures and those on the other side of the cutwater are acted upon by comparatively low pressures. The resulting unbalanced force acts toward the shaft at a point approximately 240 degrees from the cutwater, as shown in Figure 2.22.

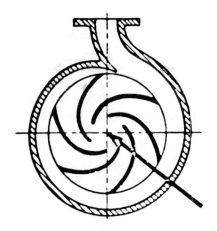

Figure 2.22. Radial loading.

At capacities greater than that of best efficiency, the resultant force acts towards the shaft at a point directly opposite.

Ina "circular" type casing, the resultant radial load is minimum at shutoff and increases to maximum near the best efficiency point. Consequently, a pump which has been selected to operate near its best efficiency point will experience much lighter radial loads if its casing is the volute type rather than the circular type.

Calculating Radial Loads

The radial load on an impeller in a volute type casing can be calculated by the use of the empirical formula:

$$P = K_Q \times K \times \frac{H \times sp. \ gr.}{2.31} \times D_2 \times B_2 \quad \text{(Equation 2.21)}$$

where:

$$K_Q = 1 - \left(\frac{Q}{Q_n}\right)^n \quad \text{(Equation 2.22)}$$

$$n = 0.7 + 2.6\frac{(N_s - 500)}{3000} \quad \text{(Equation 2.23)}$$

where:
P = resultant radial force, in pounds
K_Q = capacity factor
K = radial thrust factor at shutoff, from Figure 2.23
H = total head at Q gpm, in feet
D_2 = O.D. of impeller, inches
B_2 = width of impeller at O.D., in inches
Q = capacity, in gpm, at which radial thrust is to be calculated
Q_n = capacity, in gpm, at best efficiency of pump
N_s = specific speed of pump

A quick examination of the above equations, reveals that the radial load is greatest at shutoff, where Q is zero and $K_Q = 1.0$. As Q increases, K_Q and the radial load decreases to zero at the best efficiency point where $Q = Q_n$. As Q increases to values greater than Q_N, K_Q and the radial load increase as negative values. The negative sign indicates that the force is now in the opposite direction.

Axial Loading

Single suction impellers are subject to an axial thrust which is the net result of three forces. One force is caused by the pressure behind the impeller tending to push it towards the suction nozzle. A second force is the result of

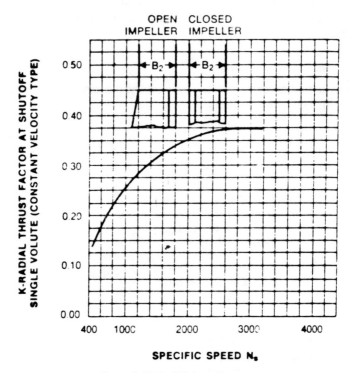

Figure 2.23. Radial thrust factor.

pressure on the other side of the impeller tending to push it toward the motor. The third is caused by the suction pressure which acts on the cross-sectional area of the shaft, as if it were a piston in the stuffing box. The pressures which act on the impeller area are created by the impeller. They are directly proportional to specific gravity of the liquid pumped.

The outboard bearing must be able to absorb this axial thrust and keep the rotating element from moving axially. The life of this bearing is, of course, depending upon the axial load, and can be greatly increased by designing the impeller so that the net axial force is very low. The formula for calculating the net axial thrust vary with each impeller style.

BEARINGS AND BEARING LIFE

Two of the most important parts in a centrifugal pump are the bearings. This may seem to be a point for discussion, but consider the vital role that they play in pump operation. They must allow the shaft to rotate with practically negligible friction, so that the horsepower requirements will be kept to a minimum. They hold the rotating element in its proper position relative to the stationary parts of the pump, both radially and axially, so that rubbing cannot occur. They must be able to absorb the forces that are transmitted to them from the impeller and give trouble free service for long periods of time.

Bearing Load

Assuming correct manufacture and maintenance, the life of a bearing is dependent upon the load that it must carry and the speed of operation. The loads on pump bearings are imposed by the radial and axial forces acting on the impeller as described.

Radial Load

Radial thrust acting on the impeller creates radial loading on both bearings. The magnitude of the radial load at each bearing can be determined by the use of the following equations:

$$R_1 = \frac{Pa}{S}$$ (Equation 2.24)

$$R_2 = \frac{P(a + s)}{s}$$ (Equation 2.25)

where:
R $\quad$ = radial load at bearings 1 and 2, in pounds
P $\quad$ = radial thrust on impeller, in pounds, from Equation 2.21
a & s = dimensions shown in the Figure 2.24.

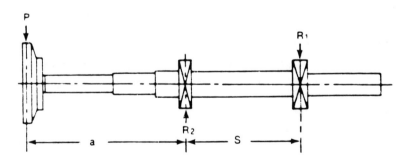

Figure 2.24. Radial thrust.

Axial Load

In any two bearing system, one of the bearings must be fixed axially, while the other is free to slide. This arrangement allows the shaft to expand or contract without imposing axial load on bearings, and yet definitely locates one end of the shaft relative to the stationary parts of the pump.

In some horizontal centrifugal pumps the outboard bearing (the closest one to the coupling) is fixed axially. The inboard bearing is free to slide within the housing bore to accommodate thermal expansion and contraction of the shaft.

Since the outboard bearing is fixed in the housing, it must carry the axial thrust in addition to radial thrust. The axial thrust is considered to be acting along the centerline of the shaft, and therefore, is the same at the outboard bearing as it is at the impeller. The radial and axial loads combine to create a resultant angular load at the outboard bearing.

Bearing Life

A bearing is capable of operating for a specific number of revolutions under a specified load, before the first evidence of failure appears. Naturally, at higher loads this number decreases, and at lighter loads it increases.

The life of an individual bearing in a specific application cannot be foretold because of unavoidable variations in bearing materials and manufacturing. However, it is possible to predict by statistical analysis that 90% of a large number of specific size and type of bearing in a specific application will last longer than a certain time. That certain time is called the "rating life" of that size and type of bearing in that specific application. The "rating life" is calculated by the following procedure outlined in ANSI Standard B3.15-1972 or the equivalent procedure usually included in each bearing manufacturer's catalog.

Bearing Lubrication

Remarks on bearings would be incomplete without comments on lubrication. Either grease or oil lubrication is satisfactory so long as it is done properly.

As a general rule, grease is preferred when:
1. Temperatures are not excessive, usually not over 200°F.
2. Speed does not exceed bearing manufacturer's recommended limit for grease lubricated bearings.
3. Extra protection is required from dirt, fumes and other contaminants.
4. Prolonged periods of operation without maintenance are expected.

Oil is preferred when:
1. Operating temperatures are consistently high.

2. Speed exceeds bearing manufacturer's limit for grease lubricated bearings.
3. Dirt conditions are not excessive and oil tight seals can be utilized.
4. Determination of the lubrication conditions is more easily detected.
5. Bearing design does not lend itself to grease lubrication.

Regardless of the type of lubricant, proper lubrication is essential if the bearing is to live up to its predicted life. Oil levels must be maintained, contamination must be prevented, regreasing must be done properly. To insure optimum lubrication, it is important that the pump user follow explicitly the instructions in the operation manual which he receives with his pump.

SHAFT SEALING DEVICES

It is highly desirable that centrifugal pumps be provided with a suitable shaft sealing arrangement so that the pumped liquid will not leak to the atmosphere. Sealing problems confronting centrifugal pump manufacturers are many because:

1. The seal is mounted on a rotating shaft which may not be running concentric, or may be worn.
2. The pressure differential may vary greatly.
3. Abrasive materials are often encountered.
4. Temperature changes are usually experienced.
5. Corrosive materials are handled frequently.

Packing

A typical pump stuffing box is shown in Figure 2.25. The packing usually consists of fibrous material that is first woven into rope and then cut and molded into packing rings of rectangular cross-section. These rings are made sufficiently soft so that pressure from the packing gland forces them to fill the space between the stuffing box and the shaft.

The packing is usually impregnated with a viscous liquid which:

1. Provides lubrication between the packing and the shaft.
2. Seals off the microscopic spaces between the packing fibres.
3. Helps to distribute heat developed by friction between the packing and the shaft.

In pumps handling corrosives, the rope packing is usually made from asbestos, tetrafluoroethylene (TFE), or graphite fiber. The choice between these three is made on the basis of cost versus corrosion resistance.

The three major virtues of a packed stuffing box are:

---

*Footnote: Additional information on the Packings and Seals is available in another separate section in this book.

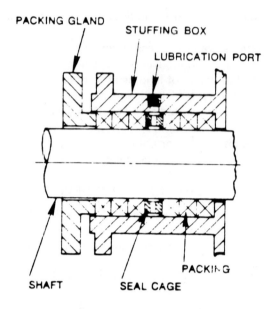

Figure 2.25. Pump stuffing box.

1. It is relatively inexpensive in initial cost.
2. It rarely is the cause of an unscheduled shutdown of the pump since its sealing efficiency usually deteriorates slowly.
3. To compensate for wear, the packing can be adjusted or completely replaced without disassembly of the pump.

The two major shortcomings of a packed stuffing box in a chemical pump are:

1. Leakage
2. The cost of packing maintenance.

The ability of rope packing to minimize leakage depends on its ability to maintain a close clearance between itself and the rotating shaft. Any radial movement of the shaft, due to runout, deflection, whip or bearing losseness, tends to open up the clearance and increase the leakage along the shaft. Every pump shaft exhibits these forms of radial movement to some degree, but in chemical pumps they tend to be accentuated by the severity of the applications.

In addition to radial movement, thermal expansion and contraction of the packing tend to open up the clearance along the shaft. Since the stuffing box temperature can change appreciably with changes in pumping conditions, this adds to the leakage problem.

Furthermore, even in the easiest of applications, it is usually necessary to allow some leakage to occur in order to remove the heat that is generated by friction between the packing and the shaft. This leakage may amount to only a few drops per minute, but in a chemical pump this may be unacceptable due to the value of the product lost or the safety hazard it may create.

While it is true that a packed stuffing box can be adjusted to minimize leakage due to corrosion, wear, and the conditions described above, the cost of labor required to make the adjustment can lead to excessively high maintenance costs if the pump has to have frequent attention.

Mechanical Seals

In most applications, a mechanical seal can be selected to overcome the shortcomings of rope packing; however, the mechanical seal will usually be higher in initial cost and presents the possibility of sudden failure, requiring shutdown of the pump for seal replacement.

A typical mechanical seal is shown in Figure 2.26. Though the details of construction may vary from one make of seal to another, the basic components are similar. Each has two mating seal rings, parts 1 and 6, held in contact by a compression spring, part 8. Ring 1 is stationary and ring 6 rotates with the shaft. There are three sealing points in the typical mechanical seal installation. One exists between ring 6 and the shaft 5. Another is between ring 1 and the end of the stuffing box. The third is normal to the pump axis, between the mating faces of rings 1 and 6.

The seal between ring 6 and the shaft is usually accomplished by one of the following means:

The rotating seal between ring 1 and 6 is achieved by lapping both mating surfaces until they are extremely flat and smooth. When these surfaces are pressed together by the spring, part 8, only a microscopically thin film of liquid exists between them, and the leakage of liquid through this gap is imperceptible.

In conventional seals, the thin film of liquid between the seal faces is the key to successful performance of the rotating seal. If the liquid film has adequate lubricating properties, the rotating seal will be maintained almost indefinitely. On the other hand, if the film of liquid contains solids, or if it heats up and loses its load carrying capacity, or if it crystallizes when the pump is shutdown, thereby cementing the seal faces together, then the seal will fail rapidly. Most of the variations in seal installations are designed to insure a continuous liquid film, with optimum lubricating qualities.

The selector of a mechanical seal for a specific application should keep in mind three factors when making his selection:

1. That the seal be made of materials that will resist the deleterious effects of the liquid being sealed.
2. That the condition of the pump be such that it will not contribute to premature seal failure due to excessive radial and axial movement of the shaft.

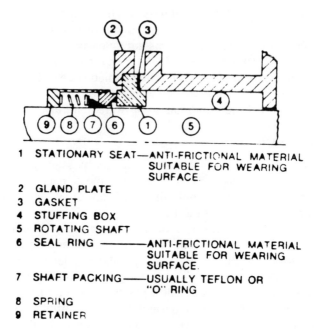

1 STATIONARY SEAT—ANTI-FRICTIONAL MATERIAL
   SUITABLE FOR WEARING
   SURFACE.
2 GLAND PLATE
3 GASKET
4 STUFFING BOX
5 ROTATING SHAFT
6 SEAL RING ————————ANTI-FRICTIONAL MATERIAL
   SUITABLE FOR WEARING
   SURFACE.
7 SHAFT PACKING——USUALLY TEFLON OR
   "O" RING
8 SPRING
9 RETAINER

Figure 2.26. Mechanical seal.

3. That the seal installation insures the continuance of a liquid lubricating film between the sealing faces.

SHAFT DEFLECTION, WHIP AND RUNOUT

As explained, radial movement of a pump shaft reduces the performance and life of its stuffing box seal. Radial movement of the shaft occurs in three forms: deflection, whip and runout.

These forms of radial movement are of vital concern to the pump user. This concern is recognized by ANSI B73.1, the Standard for centrifugal chemical pumps. It limits the radial deflection and runout of the pump shaft to a maximum of 0.002 inches at the face of the stuffing box, and calls for impeller balancing to minimize shaft whip.

Shaft Deflection

A shaft deflects, or is pushed from its normal position, because of unbalanced radial loads on the impeller. While at rest, the shaft is straight and is centered in the stuffing box. When pumping, the unbalanced radial loads make the shaft bend and take up a new position in the stuffing box.

Figure 2.27 shows a shaft in its normal position and also in an exaggerated

Figure 2.27. Shaft deflection.

deflected position. Note that the shaft revolves on its own centerline even when it is deflected, because the load is constant in direction and magnitude. In other words, it maintains its bent position as long as the speed of rotation and the quantity being pumped are not changed. Since the shaft remains in this same position as long as the capacity is not changed, and since it is revolving on its own centerline, the stuffing box packing can be adjusted to allow a minimum of leakage. When the pump is stopped or the rate of flow drastically changed, the packing may have to be readjusted to compensate for the change in deflection.

Shaft Whip

The impeller end of a shaft that is whipping revolves in such a manner as to generate a cone shape. The direction that it is moved from its true centerline changes 180° for each 180° turn of the shaft. Shaft whip makes it almost impossible to have a stuffing box function properly because the packing is pushed out to the diameter of the cone being generated. This leaves a hole which is larger than the shaft diameter and excessive leakage will occur. Usually the force that is causing whip is coming from within the rotating element. For example, an impeller head that was not properly balanced would cause whip since the heavy side of the impeller would always be on the same side of the shaft.

Figure 2.28. Shaft whip.

It is entirely possible to have a combination of shaft deflection and shaft whip present at the same time. In this case, the cone generated by the whipping shaft is moved to one side by the amount that the shaft deflects.

Shaft Runout

Shaft runout is another term that is often used in discussing pump shafts.

It is the amount that the shaft sleeve section is out of true when the shaft is in the pump. Shaft runout is measured by fastening a dial indicator to a stationary part of the pump so that its contact point indicates the radial movement of the shaft surface as the shaft is rotated slowly. A runout check is made on every pump after it is assembled. This inspection will disclose any out of roundness of the shaft, any eccentricity between the shaft and shaft sleeve, and any permanent bend in the shaft. Any of these defects would produce an oversize hole through the packing and make the stuffing box hard to seal. This check will not disclose any shaft whip that would be caused by impeller unbalance since the runout check is made by turning the shaft by hand.

VIBRATION

When we speak of vibration in a centrifugal pump, we usually refer to vibration in a radial or axial direction. This type of movement and the force accompanying it tend to damage the bearings and other parts of the pump.

It is customary to measure the vibration of a pump as close as possible to each bearing, in two radial directions (vertical and horizontal on a horizontal pump) and in the axial direction. The vibration detecting device may measure the acceleration, velocity, or the displacement of the surface which it contacts. Regardless of the parameter measured, most instruments can display all three. Centrifugal pump vibration is usually expressed in terms of peak-to-peak displacement.

Vibration may originate within a pump and motor assembly or be transmitted to it from some outside source. Table 2.2 lists the common causes of vibration. Table 2.3 defines acceptable vibration limits for centrifugal pumps.

One means to extend the usefulness of a simple vibration reading is to establish a program of measuring and recording the readings from a given pump at regularly scheduled intervals. The change in vibration, from one interval to the next is at least as important as the magnitude of the vibration. Thus, an increasing vibration level, even though it may be relatively low in value, indicates an approaching problem; whereas, an unchanging level, even though relatively high, gives evidence of a stable, and probably satisfactory, operating condition.

If the vibration of a pump exceeds the generally accepted limit noted in Table 2.3, or if successive measurements show an increase in the vibration level, it is advantageous to identify the source of vibration to facilitate corrective maintenance. One approach to identification is through vibration analysis. Basically, it involves the determination of the frequency at which the maximum vibration level occurs.

When there is a dominant frequency with no apparent relationship to pump rpm, the source of the vibration is probably external to the pump. To reduce this form of vibration, the source must be located and suppressed, or the

Table 2.2. Relationship Between Frequency and Cause of Vibration

Frequency	Amplitude	Cause	Remarks
1 x RPM	Largest in radial direction. Proportional to unbalance.	Unbalance	
1 x RPM normally	Axial direction vibration 50% or more of radial	Misalignment of coupling or bearings and bent shaft	Easily recognized by large axial vibration. Excessive flange loading can contribute to misalignment.
Very high. Several times RPM	Unsteady	Bad anti-friction bearings.	Largest high-frequency vibration near the bad bearing.
2 x RPM		Mechanical looseness.	Check grouting and bedplate bolting.
1, 2, 3, & 4 x RPM of belts.	Erratic or pulsing.	Bad drive belts.	Use strobe light to freeze faulty belt.
1 or 2 x synchronous frequency	Disappears when power is turned off.	Electrical	3600 or 7200 cps for 60 cycle current
No. of impeller vanes x RPM		Hydraulic forces	Rarely a cause of serious vibration
Extremely high, random.		Cavitation	

Table 2.3. Maximum Acceptable Vibration

Pump Speed RPM	Displacement* Mils Peak-to-Peak
3600	1.25
1800	2.50
1200	3.75

*Measured perpendicular to the shaft axis at point on the bearing housing adjacent to each bearing.

pump must be isolated from it by means of flexible connectors in the piping and/or vibration isolating mounts under the base.

HEAT GENERATED IN PUMPING

In centrifugal pumps, some of the input energy is transformed by fluid friction into heat, which increases the temperature of the liquid being pumped. The temperature increase is dependent upon the flow rate of liquid moving through the pump. At normal flow rates, this temperature rise is negligible; while under shutoff (zero flow) conditions, the temperature will continue to rise, resulting in eventual damage to the pump. It is obvious that some minimum flow rate must be maintained to prevent this from occurring. In order to determine the minimum flow rate, the maximum allowable temperature must be known.

The maximum allowable temperature rise is established by three basic considerations:

1. The temperature at which the pumped product will be adversely affected.
2. The temperature at which the properties of the pumped product are changed so as to adversely affect the action of the pump (e.g. vaporization, polymerization, etc.)
3. The maximum temperature which various components of the pump and/or piping system (e.g. mechanical seals, packing, gaskets, etc.) can accommodate.

In most applications, 10°F. is an acceptable temperature rise; however, when NPSH is critical, the temperature rise should be limited to 5°F. or less.

The minimum flow rate which must be maintained may be calculated as follows:

$$Q = \frac{5 \times BHP_o \times C_{hp}}{\Delta T \times sp\ ht} \qquad \text{(Equation 2.26)}$$

where

Q = minimum flow rate, in gpm
BHP_o = the non-viscous performance curve horsepower at shutoff
sp ht = specific heat of liquid
ΔT = maximum allowable temperature rise, degrees F.
C_{hp} = viscous horsepower correction factor (See Figure 2.16).

To find the temperature rise resulting from a known flow rate, the Equation 2.26 can be written as follows:

$$\Delta T = \frac{5 \times BHP_o \times C_{hp}}{\Delta Q \times sp\ ht} \qquad \text{(Equation 2.26a)}$$

Here is an example for calculating minimum flow rate:

Example 8

Given a 4 × 3 — 13 pump at 1750 rpm (See Figure 2.18), handling a liquid with a 1.23 specific gravity, a 0.85 specific heat, and a viscosity of 1625 centipoise, what is the minimum allowable flow rate for the 12½ ″ impeller if the temperature rise is to be kept below 10°F.?

Solution

1. From Figure 2.18, read the non-viscous shutoff horsepower for the 12½ ″ impeller equals 11.8.
2. Following the procedure outlined in the example to derive viscous performance curves (which calls for the same pump, operating under the same conditions as the example) find the viscous horsepower correction factor, $C_{hp} = 1.5$.
3. Substituting into Equation 2.26

$$Q = \frac{5 \times BHP_o \times C_{hp}}{\Delta T \times sp\ ht}$$

we find:

$$Q = \frac{5 \times 11.8 \times 1.5}{10 \times 0.85} = 10.4\ gpm$$

Minimum Flow By-Pass

In situations where the required flow rate is less than the minimum allowable flow, a by-pass line can be installed which returns a portion of the pumped liquid to the suction reservoir.

The sum of the required and by-pass flows should be equal to or greater than the calculated minimum allowable flow. The by-pass flow can be controlled with a throttling valve or an orifice to accomplish this purpose.

SELF-PRIMING CENTRIFUGAL PUMPS

A standard centrifugal pump will NOT move liquid unless it is primed. A pump is said to be "primed" when the pump casing and the suction piping are completely filled with liquid. Units that are located below the liquid level of the suction tank can be primed by merely opening the suction and discharge valves, thereby allowing liquid to flow into the pump by gravity. Standard centrifugal pumps that are located above the suction level must be primed by some auxiliary means such as a vacuum pump, or an ejector. These means of priming chemical pumps are generally unsatisfactory because of the corrosive nature of the liquids being handled. Foot valves are sometimes used on the lower end of the suction line so that it is possible to fill the pump and piping with liquid from an outside source; but this method demands an elevated tank to store the priming liquid. All of these systems require someone to be present to operate the auxiliary equipment initially, and to reprime the pump should it air-bind during operation.

The self-priming pump is one answer to the above problems. It primes itself, it will reprime if it becomes air bound during operation, and does not require constant attention during operation.

A well designed self-priming pump must be capable of efficiently removing the air from the suction piping. Removal of this air creates a partial vacuum in the suction line, allowing atmospheric pressure to force the liquid from the suction tank up into the pump, thereby establishing prime. Such a pump must be capable of forming a seal above the impeller so that atmospheric pressure cannot work back through the discharge line to fill the vacuum. Finally, a self-priming pump must function efficiently as a standard centrifugal pump after it has attained prime and established flow. For this reason, the self-priming features must not detract excessively from the pumping ability of the unit.

There are many different self-priming centrifugal pump designs, but, in essence, they all function in the same manner. That is, they all retain a certain amount of "priming liquid" when they are shut-down or lose prime, and they all recirculate the priming liquid in such a way as to entrain air at their inlet side and to release it at their discharge side.

After establishing prime, the pump performs exact as any other centrifugal pump, with the exception that it can prime itself automatically if it becomes air bound. For example: if the liquid level drops below the end of the suction pipe, the suction piping will again fill with air and pumping action will cease. As the level builds up again and covers the end of the pipe, the priming cycle will be repeated and the pump will resume normal operation.

It should be remembered that these pumps are primarily designed to convey a liquid and are, therefore, not efficient air handling machines. For this reason, particular attention should be given to the pump installation to limit the amount of air that the pump will be required to evacuate. The suction lift and the volume of the suction line should be kept to a minimum, and all the joints should be made air tight. In addition, any pressure build up at the discharge of the pump will reduce its suction lift. Therefore, the discharge line should not include check valves, closed valves, or any other obstruction that would block the flow of air and cause pressure to build.

The stuffing box area must be sealed tightly against the entrance of outside air. This is accomplished by use of quality packing or a mechanical seal. During the priming cycle particularly, the stuffing box is under vacuum and air that enters from outside must be removed in exactly the same manner as that from the suction piping. This increases the priming time; and may, if the leak is large, prevent priming altogether. Where possible, it is a good idea to provide a flushing line into the stuffing box to prevent the leakage of air into the pump at that point.

When large volumes of air have to be evacuated from the suction piping, the priming time may become so extended that the liquid recirculated within the pump becomes hot enough to vaporize. When this happens the priming action ceases and the priming chamber has to be refilled with liquid.

The maximum lift that can be tolerated depends upon the specific grafity and the vapor pressure of the liquid being handled. Heavy liquids will not be forced as high into the suction line by atmospheric pressure as will light liquids. Therefore, the maximum lift is lower for these heavy liquids. High vapor pressure liquids will tend to vaporize as they reach the areas of greatest vacuum, and the maximum lift for these liquids is therefore reduced. Both of these factors are taken into account in the calculation of the available NPSH of the system. It is more accurate to use the available NPSH to establish the maximum lift than to use an arbitrary figure. In fact, every self-priming pump installation should be thoroughly checked for available NPSH before the pump is selected.

The self-priming pump has a definite place in the pump industry, but care must be taken to insure that the pump is not misused nor expected to operate beyond its capabilities.

CHAPTER 3

POSITIVE DISPLACEMENT PUMPS

DONALD T. DEELEY
Hooker Chemical Corp.
Grand Island, NY

Positive displacement pumps trap a quantity of liquid, then force it out against the existing process pressure. These pumps fall into two classifications, rotary and reciprocating. Rotary pumps include external gear, internal gear, lobe, screw, progressive cavity, vane, flexible impeller, eccentric cam and peristaltic pumps. Reciprocating pumps include power and steam pumps, which are either the piston, plunger or diaphragm type.

In general, rotary pumps do not have check valves, produce a smooth, pulsation-free flow and can pump high viscosity fluids. However, an increase in discharge pressure can reduce pump capacity and usually only clean fluids are pumped at low to moderate pressures. Reciprocating pumps deliver at their rated capacity regardless of discharge pressure, pump slurries and abrasive solids and are used in high pressure applications. However, they require check valves and have a pulsed flow. Both rotary and reciprocating pumps can not operate against a closed discharge, and must be protected by a pressure relieving device.

This chapter discusses the operating characteristics, advantages, disadvantages, range of operation and typical applications of rotary and reciprocating pumps. Then supplementary equipment such as check valves, pulsation dampeners, drives, seals, couplings and metering controls are discussed. Finally, the parameters needed to specify a pump are described.

ROTARY PUMPS

Rotary pumps have a fixed casing enclosing either rotating gears, screws, vanes or a cam. They trap liquid between the rotor and casing and push it along the casing wall to the discharge. The most common types of rotary pumps are discussed below.

External Gear Pumps

An external gear pump has two gears, usually a drive gear and an idler gear, as shown in Figure 3.1. As the gears rotate, the unmeshing gears create a vacuum that draws liquid in between the gear teeth. The liquid is trapped between the gear teeth and the casing wall, and moved along the wall to the discharge. The intermeshing gears force liquid from between the teeth.

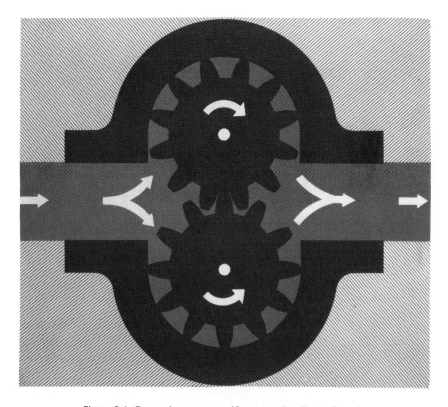

Figure 3.1. External gear pump. (Courtesy: Eco Pump Corp.)

Three gear types used are spur, helical and double helical or herringbone gears. Spur gears are used in small pumps since they are easier to machine than other gear types. Helical gears are used in dirty fluids because they are self-cleaning. Double helical gears (shown in Figure 3.2) are self-cleaning and hydraulically balanced, making them suitable for high pressure applications.

Timing gears are used in high pressure and dirty fluids applications to prevent contact between pump gear teeth. Timing gears are either internal or external to the pump casing. Internal timing gears are exposed to the liquid being pumped and are used only with clean, lubricating fluids. External timing gears are separated from the process fluid by shaft seals and are used with either dirty or non-lubricating fluids.

Advantages

Gear pumps can pump in either direction, are self-priming, produce a pulsation free flow, are lightweight, compact and low in cost. In-line gear

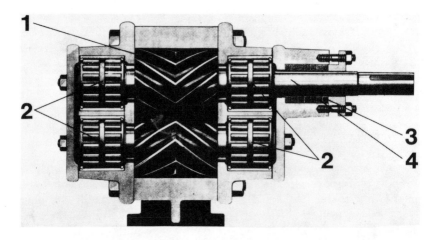

Figure 3.2. Double helical gear pump. Shown is the basic work module for types GR and GRW rotary pumps, made up of these quality components: 1. Double helical herringbone gears provide quiet, pulsation-free flow, prevent trapping of liquid. Made of high-alloy iron for maximum wear resistance. 2. Heavy-duty roller bearings assure high efficiency and long life. Manufactured by vacuum degassing process; force-feed lubricated by fluid being pumped. 3. Only one stuffing box, arranged for ample packing or mechanical seal to cut leakage. 4. Steel shaft, accurately ground by a precision-controlled method. (Courtesy: Worthington Pump Inc.)

pumps are available. In addition, gear pumps with timing gears have a long operating life and can pump liquids with small diameter solids.

Disadvantages

A pressure relief device is required to protect against operation with a closed discharge. The liquid pumped should have some lubricity and be clear, although small diameter solids can be tolerated in low concentrations. Entrained gas or vapors cause erosion. Variations in discharge pressure and viscosity cause small changes in capacity due to slip. Proper alignment is important due to the close clearances in the pump. Gear pumps require a shaft seal on the drive shaft. Externally timed pumps require seals on all rotor shafts. Internally timed and untimed gear pump bearings are lubricated by the fluid pumped. Gear pumps are noisy. A strainer should be used in the pump suction.

Range and Applications

Spur gear pumps can pump up to 45 m³/hr (200 gpm) while helical and herringbone pumps are used up to 115 m³/hr (500 gpm) although they can pump up to 1150 m³/hr (5,000 gpm). Gear pumps in general handle viscosities up to 1 × 10⁶ cs (5 million ssu), pressures up to 3450 kPa (500 psi), although 34,500 kPa (5,000 psi) are possible, temperatures up to 425°C (800°F) and speeds up to 4000 rpm. They can create a suction lift of up to 77,000 Pa (26 feet of water).

Applications include transfer pumps and lube-oil circulation on ships, metering chemical feeds, adding flocculent to water and circulating refrigerant.

Internal Gear Pumps

Internal gear pumps have a small drive gear inside a larger idler gear (See Figure 3.3). As the gears unmesh, a vacuum is created sucking liquid in between the gear teeth. The liquid is trapped between the teeth and moved along the casing wall by the outer gear and past the fixed crescent-shaped divider by the internal gear. As the gears remesh, liquid is pushed from between the external gear teeth, then squeezed from between the internal gear teeth. Timing gears are used in internal gear pumps to prevent contact between gears, although low speeds and the rolling contact between gears reduce friction, wear and turbulence in these pumps.

The advantages of an internal gear pump are the same as those of the external gear pump plus a lower shear rate. The lower shear is due to pushing liquid from between the external gear teeth instead of squeezing it from between intermeshing gears. The disadvantages are the same as external gear pumps. Internal gear pumps are used up to 250 m³/hr (1100 gpm) up to 4 × 10⁵ cs (2,000,000 ssu), up to 1700 kPa (250 psi), up to 340°C (650°F) and up to 1800 rpm. Suction lifts of 65 kPa (22 feet of water) are possible.

A typical application of internal gear pumps is pumping viscous, shear sensitive polymers.

LOBE PUMPS

A lobe pump has impellers with one, two, three or four lobes. (See Figure 3.4). The impellers must be driven by timing gears. The lobes create a suction as they both move away from the suction port, drawing liquid into the pump. Then they push the fluid along the casing wall to the discharge port.

The impellers have close wall clearances and some designs have packing strips on the lobe ends to seal along the casing wall. Adjacent lobes roll over each other to seal between impellers.

Advantages

Lobe pumps have the same advantages as external gear pumps except the flow is pulsed. Single lobe pumps have the largest pulses; two, three and four

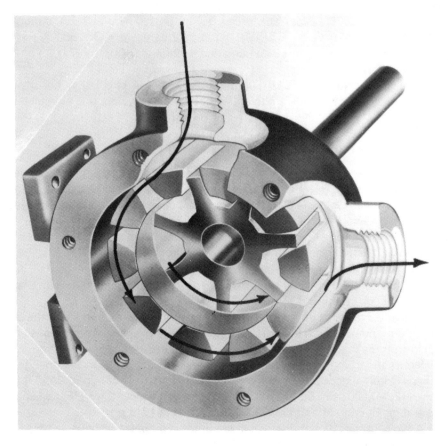

Figure 3.3. Internal gear pump. (Courtesy: Viking Pump Div., Houdaille Industries, Inc.).

lobe less. Another advantage is that alignment is less critical in a lobe pump versus gear pump. Wear is also less due to the rolling action between lobes. Lobe pumps can pump liquids with entrained gas because the rolling action between lobes provides a positive seal between the discharge and suction. The large cavities enable a lobe pump to handle liquids with large solids without damaging the solids.

Disadvantages

A lobe pump cannot pump liquids with abrasive or gritty solids due to close internal clearances. The pump has mechanical seals on both shafts and a pressure relief device is required on the discharge to protect the pump against operating with a closed discharge. Variations in viscosity and discharge pressure cause changes in capacity due to slip.

Figure 3.4. Lobe pump. (Courtesy: Mono Group Inc.)

Range and Applications

Lobe pumps range up to 450 m³/hr (2000 gpm), up to 100 Pa's (100,000 cp), up to 2760 kPa (400 psi), up to 315 °C (600 °F) and can create a suction lift of up to 60 kPa (20 feet of water). Some typical applications are pumping soups, fricasses, baby foods, and cookie batter.

SCREW PUMPS

Screw pumps include a single screw pump, progressive cavity pump, twin screw and three screw pumps.

A single screw pump has a single screw in a close fitting cylindrical casing. The pump operates the same as a screw conveyor, pushing liquid up an incline the length of the screw. A single screw pump is not self-priming and must have the first screw submerged to operate. The single screw pump is a low head pump, usually discharging against atmospheric pressure and can move liquid only the screw length. It can pump liquid with solids, rocks or wood (1). It is a low speed pump and so a low wear, low shear pump.

The single screw pump angle of inclination must be between 22° and 38°. It can pump up to 11,360 m³/hr (50,000 gpm), has up to 3m (10 feet) diameter screws, rotates between 24 and 110 rpm and can pump liquid up to 9m (30 feet) vertical height. Typical applications are pumping sewage sludge without damaging the floc and pumping irrigation water.

PROGRESSIVE CAVITY PUMP

A progressive cavity pump, sometimes called a single screw pump, has a spiral rotor turning eccentrically in an internal-helix casing. (See Figure 3.5). In the pump, cavities formed between the rotating screw and the casing move toward the discharge. At the suction, a slight vacuum is formed when a cavity is created. Liquid is sucked into the pump, then progresses toward the discharge.

Figure 3.5. Progressive cavity pump. (Courtesy: Mono Group Inc.).

There is no metal to metal contact between the rotor and casing, and so liquids with solids can be handled. Abrasive solids are pumped using a hard metal rotor and a rubber lined casing.

Advantages

A progressive cavity pump can handle abrasive, gritty solids and slurries. It can also pump semi-solids without damaging the solids (2). A progressive cavity pump does not need check valves, has a smooth discharge flow, low wear rates, a long life and can operate against a closed discharge.

Disadvantages

A progressive cavity pump is not self-priming, capacity varies with changes in viscosity and discharge pressure, and they are bulky and heavy. Seals are required on the rotating shaft.

Ranges and Applications

Progressive cavity pumps range up to 270 m³/hr (1200 gpm), up to 6200 kPa (900 psig), up to 1000 Pa's (1,000,000 cp), up to 260°C (500°F) and up to 1200 rpm.

Some applications are pumping soap, latex, resins, caustic, acids, dyes and semi-solids like potato salad.

TWIN SCREW PUMP

A twin screw pump has two intermeshing screws enclosed in a single casing. (See Figure 3.6). Liquid flows to one end of the screws and is sucked into the cavity between the screw threads as the screws unmesh. Then the liquid is conveyed along the casing wall to the discharge. This type of screw pump is used for low viscosity fluids at medium to high pressures or hydraulic fluids at very high pressures. Pressure in a screw pump increases incrementally along the pump in equal steps with each screw flight. Higher pressures are achieved in screw pumps by adding more screw flights and making the screws longer. However, since the pump is not hydraulically balanced, axial loading on the screws becomes a problem.

A similar type of twin screw pump is the duplex twin screw pump, which has two sets of opposing screws in a single casing. (See Figure 3.7). Liquid flow into the inlet port is split and flows to both ends of the screws. The opposing screws pump the liquid to a common, central discharge port. A duplex is hydraulically balanced due to the opposing flow in the pump. This balancing almost eliminates end thrust in the pump. Hydraulic balancing is used in small pumps at pressures above 1000 kPa (150 psi) and in large pumps at pressures above 340 kPa (50 psi) (3). A duplex pump is used for large volumes or very viscous liquids. For liquid viscosities above 2,000 Pa's (20,000 poise), larger

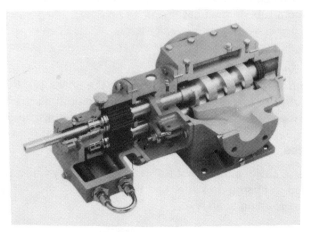

Figure 3.6. Twin screw pump. (Courtesy: Warren Pumps Inc.).

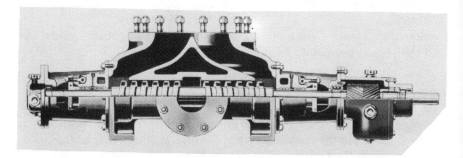

Figure 3.7. Duplex twin screw pump. (Courtesy: Warren Pumps Inc.).

than normal suction openings should be used (4). In addition, NPSH is high in duplex pumps due to reversal of flow direction at the screw suction.

Twin screw pumps are made both with and without timing gears. Untimed twin screw pumps have a drive screw and an idler screw. Another untimed screw pump is the three screw pump with a special drive screw turning two matching idler screws. Untimed pumps are less expensive than timed ones, however, metal to metal contact between the screws causes wear. In untimed pumps, the idler screw rides on the casing and so the screw must be machined smooth. Since the axial load is carried by the casing, no axial load bearings are needed.

Timed pumps have either internal or external timing gears. Internal timing gears are wetted by the fluid pumped limiting their use to clean fluids with some lubricity. Internally timed pumps are cheap and compact. External timing gears are used with non-lubricating fluids.

In duplex screw pumps, center bearings are used when sudden pressure surges will occur or when low viscosity fluids are pumped at high discharge pressures. Large diamater shafts are used when violent pressure surges can occur, including surges caused by cavitation or severe viscosity variations (5).

Advantages

Twin screw pumps are self priming, can pump in either direction, do not need check valves, have a non-pulsating discharge flow, have low wear rates and a long service life, and can pump liquids with gases and vapor. Timed screw pumps have lower wear rates than untimed pumps, and can handle non-lubricating liquids. Twin screw pumps are also low shear pumps (6). In addition, they have a large heat transfer area and a long residence time.

Disadvantages

Twin screw pumps are bulky, heavy and expensive. They can not handle liquids with solids. Their capacity changes with variations in viscosity and discharge pressure. Shaft seals are required. Externally timed pumps have more seals than internally timed or untimed types. Screw pumps must have a relief device on the discharge line.

Range and Application

Twin screw duplex pumps handle flows up to 900 m³/hr (4,000 gpm), viscosities up to 100 m²/s (5 × 10⁶ ssu), pressures up to 7,000 kPa (1,000 psi), and temperatures up to 260°C (500°F). Normal operating speeds are 1150 rpm or lower. Suction lifts up to 90 kPa (30 feet of water) suction pressure are possible.

Twin screw single inlet pumps operate with flows up to 900 m³/hr (4,000 gpm), viscosities up to 2 × 10⁵ cs (1 × 10⁶ ssu), pressures up to 9,600 kPa (1,400 psi) although 24,000 kPa (3,500 psi) are possible, temperatures up to 260°C (500°F) and speeds up to 3,600 rpm. Twin screw pumps are used in fuel oil service, hydraulics, marine cargo transfer, and pumping viscous fluids like molasses and bubble gum.

VANE PUMPS

A sliding vane pump has a slotted rotor that is eccentric to the circular casing. (See Figure 3.8). The rectangular vanes in the rotor are evenly spaced and slide freely in their slots. As the rotor turns, centrifugal force throws the vanes out against the casing. The rotor eccentricity creates a partial vacuum in the cavity between adjacent vanes. Liquid is sucked into the cavity, then pushed around to the discharge and squeezed out by the rotor. The vanes are self-

Figure 3.8. Sliding vane pump. (Courtesy: Foster Pump Works, Inc.).

compensating for wear until it becomes excessive. The self-compensation allows vane pumps to handle mildly errosive liquids. Venting is required behind the vanes to allow trapped liquid to flow out. Liquids with solids are not pumped because solids become trapped behind the vanes, plugging the vane slots. The swinging vane pump is better suited for liquids with solids. It has hinged vanes that swing out during pumping, and squeeze out the liquid trapped behind the vanes as they swing back into the rotor slots.

Sliding vane pumps vary flowrate by changing speed or changing the pump's interior eccentricity. The eccentricity is changed by rotating an eccentric casing liner while the pump is operating, changing the flowrate. The flowrate is reduced when the rotor is less off-set from the casing center, and increased when the pump interior is more eccentric. When the rotor and casing are concentric cylinders, there is no flow.

Advantages

Vane pumps are compact, lightweight, produce a non-pulsing flow, have no check valves and are self priming. Vanes and casing liners are easily replaced. Sliding vane pumps can pump fluids with viscosities lower than those pumped in gear pumps. Changes in viscosity or discharge pressure cause only small changes in capacity.

Disadvantages

Vane pumps must have a pressure relief device on the discharge and require shaft seals. A suction line strainer should be used to prevent foreign bodies from damaging the pumps. Vane pumps are not well suited for liquids with solids.

Range and Applications

Vane pumps range up to 227 m³/hr (1000 gpm), up to 1 × 10⁶ cs (5,000,000 ssu), up to 2,100 kPa (300 psi) or higher, up to 425 °C (800 °F) and from 400 to 960 rpm or higher.

Applications vary with the type of vane pumps. A sliding vane pump is used for volatile liquids. A sliding vane pump with push rods behind the vanes is used for viscous liquids. A swinging vane pump is used for non-volatile, non-viscous liquids and liquids with solids (7).

FLEXIBLE IMPELLER PUMP

A flexible impeller pump has a rotor with flexible blades turning inside an eccentric casing. As the blades unfold near the suction port and the space between the rotor and casing increases, a partial vacuum sucks liquid into the pump. The blades push the liquid around to the discharge and the casing eccentricity bends the blades, forcing the liquid out.

A flexible impeller pump should be wetted with liquid before starting the pump to lubricate the casing wall. Running the pump dry for more than a few minutes will cause heat to build up in the pump, damaging the rotor. These pumps can operate against a closed discharge, but they will be damaged by the heat build-up in the pump.

Advantages

Flexible impeller pumps are self-priming, have no flow pulsations, no check valves, are compact and lightweight, can pump in either direction and are easy to maintain. They can pump liquids with vapors and gases, liquids with solids and they are a low shear pump. These pumps can operate against a closed discharge for a short period of time without damage to the pump.

Disadvantages

Flexible impeller pumps have a limited flow range, a low pressure limit, can not pump abrasives and are not designed for heavy duty applications. They require shaft seals and must have a variable speed drive to change pumping rates.

Range and Applications

Flexible impeller pumps range up to 23 m³/hr (100 gpm), up to 1100 cs (5,000 ssu), up to 350 kPa (50 psig), up to 80 °C (180 °F) and from 75 to 1750 rpm.

ECCENTRIC CAM PUMP

In an eccentric cam pump, the cam rotates inside a circular casing. Liquid is sucked through the suction port as the cam rotates away from the port. The cavity formed is pushed along the casing wall to the discharge, where the cam pushes the liquid out. A septum between the suction and discharge ports prevents liquid from by-passing the discharge back to the suction. In an eccentric camp pump, the cam is the only moving part. In addition, there is no contact between the cam and the casing wall, reducing wear. A variation of the cam pump, the rotary plunger or cam and piston pump, substitutes a slotted vane attached to the cam for the septum separating the suction and discharge ports. The slotted vane has a longer service life than a septum.

Another type of eccentric cam pump is the flexible liner pump, with a flexible liner between the cam and the casing. (See Figure 3.9). The pumping action is the same as a cam pump, with the liquid being squeezed through the space between the liner and the casing wall by the cam. The flexible liner service life is typically one to three years, depending on the service.

Advantages

Eccentric cam pumps have characteristics of both rotary and centrifugal pumps. They are self-priming, have no check valves, have almost no pulsations in the discharge flow, are small, compact, lightweight, low shear, and can handle liquids with gases and vapors. In addition, cam pumps can operate against a closed discharge without damage to the pump.

Flexible liner pumps have all the above advantages plus there are no shaft seals, they are easy to maintain, can run dry and the fluid is isolated from the moving pump parts. Also, these pumps can pump slurries, pastes and liquids with abrasive solids.

Disadvantages

Eccentric cam pumps require shaft seals and can handle liquids with only low concentrations of solids. Flexible liner pumps do not have these disadvantages, but they will leak process fluid in the event of a liner failure.

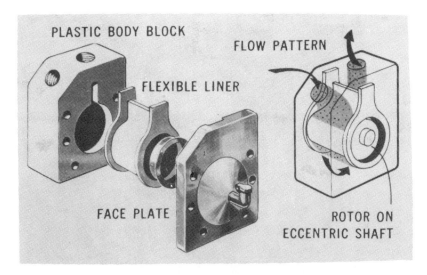

PLASTIC BODY BLOCK

FLOW PATTERN

FLEXIBLE LINER

FACE PLATE

ROTOR ON
ECCENTRIC SHAFT

Figure 3.9. Flexible liner pump. (Courtesy: Vanton Pump & Equipment Corp.).

Range and Applications

Eccentric cam pumps range up to 9 m³/hr (40 gpm), up to 1100 cs (5,000 ssu), and up to 800 kPa (100 psig). Flexible liner pumps can develop only 300 kPa (40 psig) and operate up to 120°C (250°F).

Typical application for a flexible liner pump is a transfer pump and pumping corrosives at low rates as a process pump.

PERISTALTIC PUMPS

A peristaltic or flexible tube pump has an impeller with rollers on the ends and tube inside a casing. (See Figure 3.10). As the rotor turns, rollers press against and run along the tubing squeezing the liquid through the tube, producing flow. The tube must be resilient enough to allow the rollers to collapse it and then spring back to its original shape to create a suction. Tubing sizes typically range from 0.5 to 1.25 cm (3/16 to ½ inch).

Advantages and Disadvantages

A peristaltic pump has no check valves, no shaft seals and the process fluid is completely isolated from the pump. They can handle liquids with gases and vapors, are low shear, self-priming, small, compact, lightweight, can run dry and produce a non-pulsed flow.

Disadvantages include limited pressure, low flow range and short tubing life. Flow rate is changed by varying pump speed.

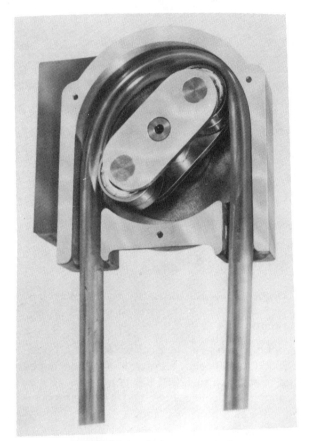

Figure 3.10. Peristaltic Pump. (Courtesy: Little Giant Pump Co.).

Range and Applications

Peristaltic pumps range up to 0.1 m³/h (0.5 gpm) and up to 170 kPa (25 psig). They can pump any flowable liquid. The temperature range depends on the tubing material selected. Typical applications are metering corrosives in laboratories and pumping blood plasma in hospitals.

RECIPROCATING PUMPS

Reciprocating pumps are classified as either power or steam pumps. Power pumps have either a motor, engine or turbine driving a cam which is connected to a reciprocating shaft. Steam pumps, also called direct acting pumps, have a reciprocating piston driven by steam or compressed gas. The piston drives the reciprocating shaft. One exception is the direct acting diaphragm pump where compressed gas drives a flexible diaphragm, not a piston.

POWER PUMPS

Power pumps are economical to operate due to their high pump efficiencies and long service life. (See Figure 3.11). Another advantage is constant

Figure 3.11. Power pump. (Courtesy: Union Pump Company).

discharge capacity against changes in discharge pressure. Power pumps will deliver at their rated capacity regardless of discharge pressure up to either the relief valve setting or their mechanical limits. This advantage is important in metering pump applications.

A major disadvantage of power pumps is size and weight since they are bulky and heavy. A second disadvantage is pulsed flow. This problem is common to all reciprocating pumps, but is more severe in power pumps than in steam pumps. Pulsed flow is discussed more in the following section on steam pumps.

Power pumps operate from 0.4 1/h to 90 m³/h (0.1 gph to 400 gpm), up to 6,000 rpm although 300 to 800 rpm is the common operating range, and up to 1.5 × 10⁶ W (2,000 Hp). Power pumps can pump at up to 3.5 × 10⁵ kPa (50,000 psig), up to 500 strokes per minute and up to 260°C (500°F).

Typical applications are supplying water for high pressure water cleaning jets and pumping grout into foundations. Metering pumps are a common power pump application and are discussed in the Miscellaneous Accessories — Metering Control section.

STEAM PUMPS

Steam pumps have three major parts on the drive end; the piston, cylinder and the steam valve. (See Figure 3.12). The piston can be one of the three types: the body and follower type where the piston has either fiber or soft metal rings, the solid type with hard metal rings, or the cup type. The body and follower type and cup type pistons are used in low pressure applications. The solid type piston is used with high pressures and high temperatures. The cylinder is usually fitted with a replacable liner due to wear caused by the piston. Wear can be reduced by adding a lubricant to the driving compressed gas. However, this is usually not feasible when general plant steam is used which can not be contaminated with oil.

The steam valve is one of the two types: unbalance or balanced. The D-slide valve (Figure 3.12) is an example of an unbalanced valve and is commonly used in low pressure applications (below 1480 kPa, 200 psig) (8). High pressures would cause excessive wear and gauling at the valve port openings. A balanced valve (Figure 3.13) is used in high pressure, high speed applications.

Both the D-slide valve and balanced steam valve operate in a similar manner. Tie rods connected to the reciprocating drive shaft cause the valve to slide back and forth, opening first the inlet port on the left, then the inlet on the right. The valve always opens the exhaust port ahead of the moving piston. When the piston passes the exhaust port, a cushion of steam is trapped in the cylinder, preventing the piston from hitting the cylinder end.

The advantages of steam pumps are smaller size, lower cost and higher

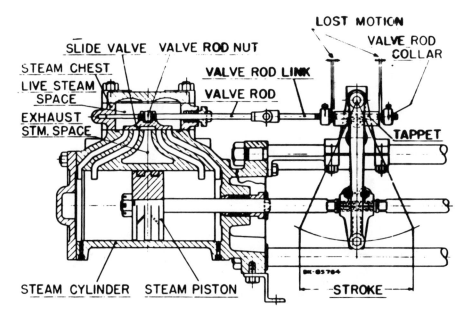

Figure 3.12. Steam pump with a D-slide valve. (Courtesy Worthington Pump Inc.).

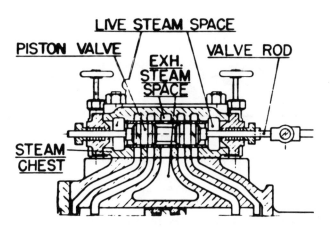

*Figure 3.13. Cross section balancd steam valve. (Courtesy
Worthington Pump Inc.).*

reliability than power pumps. A fourth advantage is smaller pulses in flow than
a power pump. A comparison of flow patterns is shown in Figure 3.14. The
Figure 3.14a represents the discharge from a duplex steam pump, while the

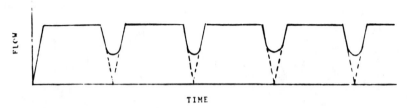

Figure 3.14a. Duplex steam pump

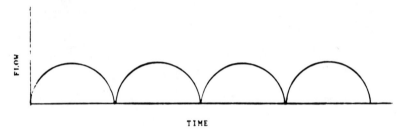

Figure 3.14b. Duplex power pump.

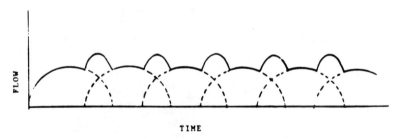

Figure 3.14c. Triplex power pump.

Figure 3.14. Flow characteristics of steam and power pumps.

Figure 3.14b represents a duplex power pump. Although both flows are pulsed, the surges are more severe in a power pump than in a steam pump. In order to reduce the amplitude of the pulses, power pumps usually have multiple heads, such as triplex or quadriplex types. The discharge from a triplex is shown in Figure 3.14c. However, unlike power pumps, adding more heads to a steam pump merely increases flow, and does not reduce the pulse amplitude. With any reciprocating pump, pulses can be reduced with a pulsation dampener. Other advantages of steam pumps are they can operate against a closed discharge without damage to the pump and pump capacity is easily varied by changing the steam flow to the pump.

A major disadvantage is that flow varies with discharge pressure. Flow decreases with increases in discharge pressure. A second disadvantage is high steam usage and so high operating costs. This cost can be offset by using the pump exhaust steam for process heating applications, since only the pressure energy, not the heating value of the steam is used in the pump. A third disadvantage is wear on the steam side, both in the cylinder and the steam valve. When compressed gas is used, wear can be reduced by adding a lubricant to the gas.

Steam pumps operate up to 7,000 kPa (1000 psig) discharge pressure with up to 1800 kPa (260 psig) steam pressure. Maximum process temperature is 177 °C (350 °F). Steam pumps should not be operated at stroke speeds slower than 2.5 cm/s (5 feet per minute). Steam pumps are used for boiler feed applications and general service on board ships.

RECIPROCATING PUMP HEADS

Both steam and power pumps have either piston, plunger, or diaphragm pump heads. A piston pump has a piston moving back and forth in a cylinder and can be either single acting or double acting. The piston can be one of the three types described in the steam pumps section of this chapter.

Two advantages of a piston pump are large capacity and complete emptying of the cylinder on the discharge stroke. These qualities are useful when pumping volatile liquids. Special check valve designs allow complete emptying of the cylinder. They prevent volatile liquid from being trapped in the cylinder and flashing on the suction stroke causing vapor binding in the pump. A common application is pumping propane (9).

Disadvantages of the piston pump are pressure limitations and leakage around the piston seals. Leakage around the piston seals goes undetected until the pump capacity drops significantly. Then the piston packing can not be adjusted without disassembling the pump. For this reason, fluids with some lubricity are pumped in piston pumps. Piston pumps range from 0.7 to 91 m³/h (3 to 400 gpm), up to 2860 kPa (400 psig), up to 177 °C (350 °F) and from 20 to 33 strokes per minute.

A plunger pump has a packed plunger moving in a cylinder. A single cylinder plunger pump has two plungers moving in a single cylinder whereas an individual cylinder plunger pump has a single plunger in a single cylinder. A plunger is usually solid in sizes up to twelve centimeters (5 in.) in diameter and hollow in larger diameters. At pressures above 40,000 kPa (6000 psig) a solid plunger may buckle and the manufacturer should be consulted (10). Plunger pumps are usually vertically mounted to prevent uneven packing wear.

One advantage of a plunger pump is the packing can be adjusted while the pump is operating. Another advantage is the packing can be flushed with a clean fluid for pumping abrasive slurries at high pressures.

A disadvantage of plunger pumps is pulsed flow. This is reduced by using multiple heads with as many as nine heads together. The pulsations are also smaller when an odd number of heads are used such as a triplex versus a quadruplex pump (11). Another disadvantage is that leakage around the packing makes these pumps less suitable for metering applications. However, controlled leakage rates allow their use in chemical feed applications when the leakage loss is low compared to the overall pump capacity, and when the leakage rate is constant. A third disadvantage is the cylinder is not completely emptied on the discharge stroke, making the pumping of volatile liquids or gases impractical due to vapor binding.

Plunger pumps operate at up to 90 m³/hr (400 gpm) and higher, up to 345,000 kPa (50,000 psig), up to 400°C (750°F) and usually between 45 to 90 meters per minute (150 to 350 feet per minute) for longer packing life.

A special multiple plunger pump is the wobble plate pump with a swash plate drive. The pump has several plungers and cylinders arranged in a circle. An undulating wobble plate drives the plungers in and out, creating a pumping motion (12). This pump has smaller pulsations than the packed plunger, but is limited to clean fluids, preferably self-lubricating liquids due to the metal to metal contact between the plunger and cylinder. A common application is pumping hydraulic fluids, since the pump is compact, and can deliver a small amount of fluid at high pressure.

DIAPHRAGM PUMP

An example of a power diaphragm pump is shown in Figure 3.15, a metering pump. Power diaphragm pumps have either mechanically or hydraulically driven diaphragms. In the first case, the diaphragm is connected directly to the reciprocating drive shaft. For high pressure applications the diaphragm is reinforced with concentric metal rings. In hydraulically driven diaphragm pumps, either a plunger or piston pushes oil against the diaphragm, causing the pumping action. On the return stroke, the plunger creates a vacuum filled by the oil and the returning diaphragm. The vacuum is limited by a vacuum breaker built into the pump housing. To avoid overheating the oil, a portion of oil is recirculated back to a reservoir on each stroke.

The main advantage of diaphragm pumps is separation of the process fluid from the critical pump parts. Only the diaphragm, check valves and pump head are wetted by the process fluid, making these pumps well suited for corrosive liquids and abrasive slurries. A second advantage is zero leakage, in theory, which is important in metering applications. Leakage only occurs around the check valves, not the diaphragm.

One disadvantage is that the number of suitable diaphragm materials is

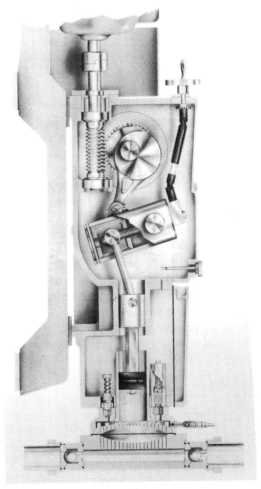

Figure 3.15. Power diaphragm pump. (Courtesy: Interpace Corp.).

limited. Normally, rubber materials are used, creating a temperature limitation. Metal diaphragms are available for high temperature and pressure applications, but metal fatigue is a problem. A second disadvantage is that diaphragm failure is catastrophic when it occurs. This should be kept in mind when selecting a hydraulic pumping oil. The oil should be compatible with the process fluid to avoid violent reaction in the event of diaphram leaks. In addition, the pump layout should take into account the possibility of diaphragm failure and provide accessibility for diaphragm service. The manufacturer should be consulted as to expected life with the process fluid and diaphgram selected, and a regular replacement schedule established.

Power diaphragm pumps are commonly used as metering or proportioning pumps. They operate from 3 to 5,500 l/h (1 to 1500 gph), up to 34,600 kPa (5,000 psig) pressure, and up to 250°C (500°F) or higher. A low stroke rate promotes longer diaphragm life, but a high stroke rate reduces pulse amplitudes.

Typical applications are chemical feeds, acid-base addition for pH control, and flocculant addition in water treatment.

Direct acting diaphragm pumps have the same advantages as power diaphragm pumps. However, they are not well suited for metering applications because, as with steam pumps, flow varies with process pressure changes. A common type of direct acting diaphragm pump is the pneumatic diaphragm pump where compressed gas is used to push the diaphragm. A duplex pneumatic diaphragm pump is shown in Figure 3.16. The slide valve alternately opens the air inlet and exhaust ports to each diaphragm. Since the diaphragms are connected, when one is pushed out by compressed air the other is forced to move in and exhaust air.

Pneumatic diaphragm pumps are designed to handle slurries. They have large inlet and outlet openings and the diaphragm moves at a high velocity keeping the fluid flowing at a high velocity, preventing settling out of solids. These pumps have all the advantages of steam pumps, such as compact size, low cost and ability to operate against a closed discharge without damaging the pump.

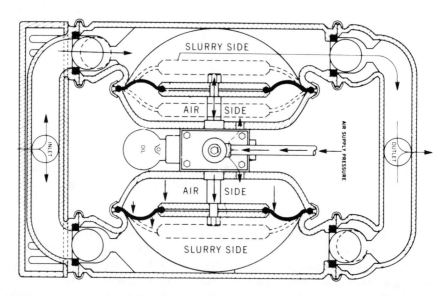

Figure 3.16. Duplex pneumatic diaphragm pump. (Courtesy: Wilden Pump & Engineering Co.).

An important problem is pulsation. The high diaphragm velocity creates large pulses which can damage both suction and discharge piping. A pulsation dampener should always be used. These pumps operate at up to 45 m³/h (200 gpm), up to 800 kPa (100 psig), and up to 150 °C (300 °F). Typical applications are pumping sewage sludge and abrasive solids.

MISCELLANEOUS ACCESSORIES

Both rotary and reciprocating pumps have accessories that must be specified when ordering a pump. These auxiliary items are discussed briefly in the following section.

Check Valves

Reciprocating pumps require check valves in order to operate. Six types commonly used are disk or flat plate, wing valve, bevel faced, ball, plug and slurry valve. A disk is used on clean fluids up to 34,000 kPa (5,000 psig). A wing valve is used with clean fluids for pressures up to 69,000 kPa (10,000 psig). Both bevel faced and ball valves are used for fluids with particles or clean fluids at pressures up to 200,000 kPa (30,000 psig). A double ball check valve is used for gritty slurries. A plug type check valve has the same applications as a disk valve. A slurry valve has either a replaceable elastomer insert or elastomer seat and is used with slurries up to 17,000 kPa (2500 psig) (13). Generally, the pump manufacturer recommendation for check valves can be followed after giving him the process fluid properties and pump operating conditions.

Clean liquid velocity through the suction port check valves is usually 1 to 2.4 m/s (3 to 8 fps) and through the discharge is 2 to 3.7 m/s (6 to 12 fps). In slurry applications, 2 to 3.7 m/s (6 to 12 fps) is used in both the suction and discharge valves. Suction and discharge valve manifolds have velocities at least as high as those used for the pump check valve ports (14).

Back Pressure Valves

A back pressure valve maintains enough pressure in the discharge line to shut the discharge check valve properly. A back pressure valve either restricts flow or acts as a pressure relief valve to maintain pressure at the pump discharge. When specifying a pump, give the vendor the minimum differential pressure expected in order to aid his selection of a back pressure valve. Valves that frequently open and close may cause a problem either in the pump or the discharge piping.

A back pressure valve should not be used as a check valve to prevent syphoning since valve will respond to pressure at either port (15). The pump discharge check valve should not be used as a back pressure valve even if it is spring loaded. The check valve spring is not powerful enough to tightly close the valve. A back pressure valve should not be used with a slurry since it will plug.

Relief Devices

Most positive displacement pumps require a pressure relief device to protect against overpressure. External relief valves are usually used because it is easy to see when they open, they can be easily set at any pressure, they remain open only long enough to relieve pressure and so reduce process fluid loss and they dissipate pump heat if the valve discharge is not piped directly to the pump suction. Internal relief valves are commonly installed in hydraulically driven diaphragm pumps as protection in case the external relief fails to work. A major drawback is internal relief does not dissipate heat. A rupture disk is usually used with slurries since an external relief valve would plug.

An external reflied device should be located as close as possible to the pump discharge, and before any back pressure valve.

In a duplex recriprocating pump, set the relief valve at 25% above the maximum operating pressure. In a triplex, quadrex, etc. pump, set the relief at 10% above maximum discharge pressure (16). In rotary pumps, set the relief at 10% above maximum discharge pressure (17).

Pulsation Dampeners

Pulsation dampeners, also called accumulators, are used to reduce pipe movement, hydraulic noise and pump discharge pulsations. Three common types are the standpipe, bladder and in-line accumulator. The standpipe is a chamber that tees into the process line. It is at least the same diameter as that line and is eight to ten pipe diameters high (or at least 0.6 m (2 feet)) with a 30 cm air cushion and a connection on top for recharging air (18). A bladder is similar to a standpipe, except a flexible diaphragm separates the air cushion from the process fluid. An in-line accumulator is an enlarged section of pipe with baffles. Standpipes and bladders are used on both suction and discharge lines, whereas an in-line accumulator is used only on discharge lines.

A pulsation dampener should be used on the discharge when both the pump head and discharge piping are metal and the flow is greater than 450 l/h (120 gph) per pump head. Always use a dampener with plastic pipe and with plastic pump heads unless the discharge line is less than 9 meters (30 ft) long and it discharges to atmospheric pressure. A pulsation dampener is used on suction lines longer than 3 meters (10 feet) (19).

An in-lne accumulator is usually used with corrosives. A bladder, when used on the discharge line, is pressurized to 66% of the system discharge pressure (20). A standpipe on the suction line can also be used to calibrate the capacity of a small pump when it is used as a small feed tank.

Bearings and Lubrication

Bearings are used to carry both radial and axial thrust loads. When specifying a pump, the manufacturer's recommendation on bearings is usually followed. A water cooled jacket is used when bearing temperatures exceed 120°C (250°F).

When using a V-belt, the overhung load on the pump and motor bearings should be checked, especially at low speeds and high horsepower requirements, to insure the load does not exceed permissible levels. Using a timing belt or special bearing arrangements help reduce the load.

Bearings can be lubricated either by the process fluid, by grease or oil. The process fluid is a good lubricant if it can coat metal parts and reduce metal to metal contact. It should normally have a viscosity of at least 7 centistokes and be non-corrosive. The fluid must circulate to remove friction heat and most pumps have small holes drilled in the pump casing to bleed liquid past the bearings. Examples of good lubricants are glue, molasses, paint, asphalt and chocolate.

Grease lubricants are used in lightly loaded pumps such as general and chemical process services. An oil lubrication system is used for medium to heavy duty applications such as refinery service.

Shaft Seals

The rotating shaft of a rotary pump can be sealed by packing or a mechanical seal. Packing is a pliable or semipliable material forced into the stuffing box and against the shaft by a movable collar, the gland. Packing must leak liquid to prevent burning and scortching the packing, since smooth, scortched packing does not seal. Packing should leak about 40 to 60 drops per minute (21). A lantern ring is often placed in the middle of the packing so that lubricant can be injected into the packing. In slurry pumps, a flush liquid is injected at the lantern ring to keep solids away from the packing, preventing scoring of the shaft sleeve and excessive leakage. A flow of 45 to 115 l/h (12 to 30 gph) is typical. The flush fluid line pressure should be between 70 and 100 kPa (10 and 15 psi) above the stuffing box pressure (22).

Besides rotating shafts, packing also seals around plungers and reciprocating shafts in reciprocating pumps. The leakage rate for these applications is the same as a packed rotating shaft. If the stuffing box is under negative pressure, a flush fluid is used.

Shaft sleeves are usually used with packing to protect the shaft. In corrosive liquid applications, a corrosion resistant sleeve is used.

Mechanical seals, which are more common than packing, are either the shaft packing or bellows type. In the packing type, the packing around the shaft slides along the shaft as the rotating seal face wears. With the bellows type, the end of the bellows is fixed to the shaft, forming the shaft seal. All mechanical seals have a stationary seal ring and a rotating face held against the seal ring by a spring. Mechanical seals are usually used on clean fluids. For liquids with solids, a flush fluid is used to keep solids away from the seal. For highly toxic fluids or gases and vapors, a double mechanical seal is used with a compatible lubricating flush fluid flowing between the two seals. In high temperature applications ($>260\,^{\circ}$C) use a water jacketed stuffing box.

Alignment

There are two types of misalignment — angular and parallel. In angular misalignment, the two shaft centerlines are at an angle to each other. With parallel misalignment, the centerlines are parallel but are offset from each other. Shafts can have both angular and parallel misalignment.

The alignment of shafts connected by a flexible coupling is checked with feeler or taper gauges at four points, every 90 degrees around the shafts. The alignment tolerance depends on shaft speed, shaft length, and severity of service. High speeds, long shafts and heavy shaft loading require higher than normal alignment precision. Maximum allowable angular misalignment is normally about 1 degree (23). In addition to alignment, the gap between shafts should be set per the coupling manufacturer's recommendation. After the shafts are coupled, they should be turned by hand to see if they move freely without binding.

Belt drive alignment is checked by putting a straight edge on the two wheel rims of the pulley and squaring the rims to the straight edge.

The pump and drive shafts should be aligned prior to connecting the pump suction and discharge ports to the process piping. Then the alignment should be checked after the connections are complete and, if possible, with the pump at its operating temperature. The alignment of in-line pumps where the drive is part of the pump, should be checked after installation.

Pump shaft alignment is adjusted by putting metal shims under the pump base, and by sliding the pump and motor in slotted bolt holes. When positioning steam pumps, the drive end should be left free to slide due to thermal expansion of the steam chest.

Slight misalignment can cause many problems. Among them are excessive vibration, wasted power and damage to both the drive and pump. Bearings and gears in the drive can be damaged by misalignment. Shaft seals in the pump can wear excessively and the gears in a gear pump may seize due to misalignment. Pumps can become misaligned while in service. Their alignment should be checked occasionally, especially with gear and screw pumps.

Couplings

Couplings reduce the effects of misalignment between the pump and drive shafts. Couplings are usually either single or double engagement type. A single engagement type, such as a universal joint, only compensates for angular misalignment. Double engagement types allow for both angular and parallel misalignment. Examples are jaw, chain, biscuit, bellows, gear sleeve on geared hubs, flexible central disks, radial-slip clutches and cushioned pins.

The flexible couplings above are used when the pump and drive each have independent bearings. A rigid coupling is used when the bearings are only in the driver, such as in a vertical pump. Generally, the manufacturer's recommendation is followed in selecting a coupling. In addition, common plant practice should be considered when selecting a coupling. Once a coupling is selected, the shaft torque should be calculated so that the coupling torque limit is not exceeded.

Motors and Drives

Air motors, steam turbines, electric motors and internal combustion engines are used to drive rotary pumps. Usually electric motors or steam turbines are used, with electric motors being more popular. (Information on drives is available in a separate chapter in this volume.)

When selecting a motor, calculate the speed and horsepower needed. Then select the next lower speed and higher horsepower standard motor unless higher speeds will be needed. Consider the starting load, fluid properties at rest and process conditions when calculating the horsepower. The motor must develop the horsepower required at the maximum possible discharge pressure, which is the relief pressure. Induction type electric motors are used up to 3.7×10^5 watts (500 horsepower). The motor enclosure type is determined by the plant hazard rating class. A gear motor is used up to 3.7×10^4 watts (50 horsepower) unless easy access to the pump is required (24). In that case, a separate gear reducer is used.

Instead of gear reducers, rotary pumps often have belt or chain drives due to the wide range of speeds required. V-belts are used when an economical drive is needed, and especially when the pump speed will be changed in the future. The overhung load of a belt can be carried by an extra pump shaft bearing. The distance between the motor and drive wheel should be kept to a minimum to reduce the overhung load.

Variable speed units are used to change speed while the pump is running to increase the pumping range. Generally, variable speed units are bulky, heavy and expensive. Examples are belt type, traction type, electronic variable speed motor, hydraulic drive and air motor.

Strainers

A strainer is used whenever possible to keep large foreign particles out of the pump. The strainer area is usually three to four times the pipe sectional area. Consult the pump vendor for a recommended screen mesh size. Coarse strainers are used for slurries and oversized strainers are used with viscous fluids. Removable screens are used during startups (25).

Pressure Gauges

Pressure gauges are mounted on the suction and discharge piping to show process conditions, to indicate flow and to show changes in pump condition. They are also useful during pump start-up for diagnosing pump problems.

Spare Parts

The pump manufacturer literature usually has a list of selected spare parts. Keep on hand all wearing parts, such as seals, shaft sleeves, wear plates, piston rings, and valve parts. When the pump vendor warehouse is nearby, rely on this warehouse as inventory for parts not likely to fail. When buying a new pump, consider relying on interchangeable parts already on hand for existing pumps. For critical pump applications where a pump failure would result in an expensive production outage, have an installed spare (26).

Base Plate

The base plate is supplied by the vendor and is either steel or cast iron. The base plate serves three functions: aligns the pump and drive; allows for easier handling of the pump; and accommodates differences in height between the pump and drive shafts. The base plate also acts as a drip pan and can be fitted with a small drip pan to catch seal and oil leaks.

The pump base plate is usually mounted on a concrete foundation by bolting the base plate to bolts anchored in the concrete. Keep in mind that the base plate flexes with the tightening of the bolts. After the base plate is bolted down, the cavity underneath the base plate is filled with grout.

Piping

The optimum piping size should be selected based on power cost, pump size and head loss. However, the suction pipeline should not have a diameter less than that of the pump section.

Piping strain on the pump should be avoided. Both the suction and discharge piping should be supported to avoid distorting the pump casing. With reciprocating pumps, suction and discharge piping should line up without forcing the flange bolts. Otherwise, pump misalignment could occur or a pump flange could crack. An expansion joint should be used when thermal expansion will be greater than 0.025 to 0.05 centimeters (27). If pipe strain already exists, a piece of reinforced rubber hose or metal hose can be used at the pump flange to reduce strain on the pump.

Self priming pumps used for lift applications should have short, straight suction lines. The suction pipe should be at least the same pipe size as the pump suction, and preferably one or two sizes larger. The piping should be air tight to insure maximum lift and to minimize priming time. Usually a foot valve is installed at the suction line entrance to avoid losing the pump prime. Since most rotary pumps operate better when wetted with liquid, a priming port should be included in the piping layout. More detailed recommendations on suction piping are given in the Hydraulic Institute Standards (28).

Metering Controls

Rotary pumps, in general, vary capacity by changing shaft speeds. Speeds can be varied with either a variable speed motor or a variable speed drive. Both of these are discussed in separate sections on drives and motors. A sliding vane pump can also vary capacity by changing the eccentricity of the pump casing. This is done by rotating an eccentric casing liner in the pump. A more detailed description of this can be found in the section on vane pumps.

One advantage of speed variation is that it is more economical than stroke length controls on multiple head pumps. Only one speed control is required versus individual stroke control for each head. Another plus is wider range of flow adjustment. A variable speed reducer commonly has a range of 10:1 for each speed. However, by changing the motor or drive gear, this range is expanded. Typically, motor and gear combinations range over speed ratios of 25:1 in addition to the range of the reducer.

Two disadvantages of speed control are poor accuracy and, in manual applications, a tackometer is needed to set the speed desired.

Reciprocating pumps vary capacity by changing the speed or changing stroke length. Stroke length is changed either by a lost motion type control or by an amplitude modulation mechanism.

Lost motion controls are used in hydraulically accuated diaphragm pumps. With these controls, the plunger stroke length remains constant. Pump capacity is regulated by controlling the location of a by-pass valve on the plunger cylinder. At 100% flow, the plunger forces oil against the diaphragm on its entire stroke. At less than maximum flow, say 50%, the oil is pumped on half its stroke to the diaphragm, and through the by-pass on the remaining half stroke. By varying the location of the by-pass valve opening, the pump capacity is varied. The main advantage of this control is that it is inexpensive compared to amplitude modulation. The main disadvantage is highly pulsed flow and the resulting shock loads on the discharge piping and pump drive. Figure 3.17 shows the output characteristics of a lost motion pump. At maximum capacity, the flow rate is the same as in any power pump. However, at 50% of capacity, the flow is highly pulsed. It drops from maximum flow to zero flow quarterway through the pumping cycle. The sudden loss of flow

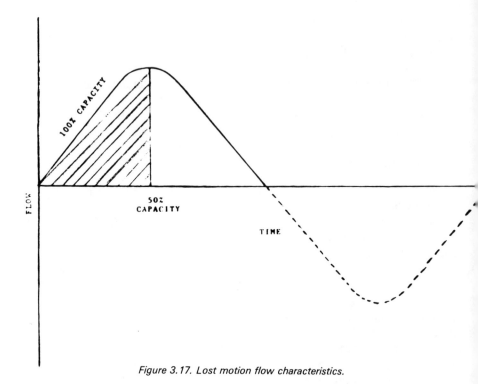

Figure 3.17. Lost motion flow characteristics.

causes several pulsations in the discharge line. The same pulse occurs a half-cycle later in the suction line.

An amplitude modulation mechanism varies the stroke length to change pump capacity. The stroke length can be varied in several ways. One is shown in Figure 3.15, where the adjusting screw on the rocker arm changes the shaft length by changing the rocker lever length between the two shaft ends. Maximum flow occurs when the shaft ends are together (minimum lever length) and minimum flow occurs when the adjustable shaft end is at the center of oscillation.

Another way to change stroke length is to vary the angle of a wobble plate that drives the connecting rod. A similar way is to vary the eccentricity of a drive gear. All the above ways are used in metering pumps.

The advantage of an adjustable modulation mechanism is pump output always follows a sine wave pattern, as shown in Figure 3.18. At maximum flow the flow characteristic is the same as a lost motion mechanism. However, at 50% maximum flow, the flow characteristic of the amplitude modulation mechanism is still a sine wave whereas the lost motion flow pattern becomes choppy and highly pulsed. The smaller pulses in amplitude modulation produce a smooth, more even flow. Also, in multiple head applications,

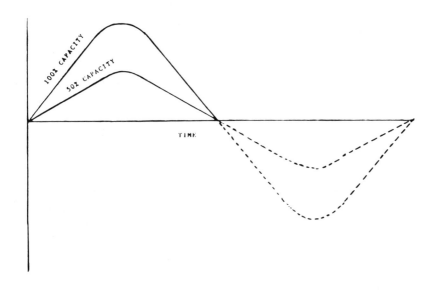

Figure 3.18. Amplitude modulation control flow characteristics.

amplitude modulation continues to have a smooth flow at less than 100% capacity, whereas lost motion becomes choppy even with multiple heads.

The major disadvantage of amplitude modulation devices is higher cost compared to lost motion control.

Both types of controls can be operated either manually, by pneumatic or electric controls. Changing stroke length maintains a high process fluid velocity which causes good fluid mixing and prevents settling of solids. Lost motion controls provide higher fluid velocities than amplitude modulation.

Varying stroke length is more accurate than changing speed, although changing speed can be cheaper.

Pump Specifications

When specifying a pump, the vendor should be given the following information:

- The type of pump — simplex, duplex, plunger, piston, gear, eccentric cam, etc.
- Fluid characteristics — % solids, % gas in liquid, corrosive compounds present, type of solids (abrasive, gritty, large, small) fluid lubricity.
- Fluid temperature — at normal and maximum conditions.
- Fluid properties — specific gravity, viscosity and vapor pressure at two temperatures.
- Flowrate —normal, maximum and minimum.
- Net positive suction head available.
- Differential pressure across the pump — normal and minimum.
- Maximum discharge pressure.
- Materials of construction.
- Type of drive and driver — motor type, coupling type, speed, type of speed reducer.
- Type of capacity controls (if any).
- Type of service — continuous, 8 to 24 hours/day; light, 3 to 8 hours/-day; intermittent, up to 3 hrs/day total or cyclic service.

In addition to the above information, pump industry classifications should be considered as well as any regulations governing pump construction and materials. Examples are sanitary pumps, fire fighting equipment, military specifications and OSHA regulations.

The following sections describe in detail some of the items listed above.

Materials of Construction

Selection of the pump material of construction is based on corrosion rate and operating temperature. A corrosion rate of a few thousandths of an inch per year may greatly increase slip in the pump, thereby reducing capacity. Corrosion rates increase with increasing flow rates. It follows that highly corrosion resistant pumps should be used in high flow applications. In addition, intermittent service pumps have higher corrosion rates than continuous service pumps.

Temperature is important in rotary pumps with close clearances. These pumps should have materials with compatible coefficients of thermal expansion. In cases of large temperature differences, such as from a cold start-up to a high normal operating temperature, materials with low coefficients of thermal expansion should be used. A steam jacket and a gradual process warm up helps during cold start-ups.

Rotary pumps are usually made with bronze or cast iron casings and cast iron or steel rotors. At temperatures above 230°C (450°F), forged or cast-steel housings are used. Special materials, such as stainless steel, nickel alloys, Hastelloy and others are available. Casing liners are available in all the above materials. Twin screw pump rotors sometimes have hard coatings to reduce wear. Common coating materials are tungsten carbide, chrome oxide and ceramics. Galling can occur in twin screw pumps with certain materials combinations. Screw and casing combinations that have a long wear life are cast iron screws in a steel or chrome casing, bronze screws in steel, monel in chrome and hardface in chrome casings (29).

Progressive cavity pumps have casing liners made of rubber or Neoprene* for abrasive slurry pumping applications. Natural rubber is the best material for this application with Neoprene* used when resistance to oil is required.

Peristaltic pumps commonly have gum rubber, Neoprene* or clear plastic tubing inside. Flexible liner and flexible vane pumps usually use rubber or Neoprene* internal parts with Viton* and Teflon* coated rubber available.

Reciprocating pumps come in three standard types — all iron, bronze fitted and all bronze. In all iron pumps, every part that contacts the process liquid is made of ferrous metal. Steel, ductile iron and 316 stainless steel are common. In bronze fitted pumps, the casing is ferrous metal, wearing parts are bronze and the shaft is either steel or nonferrous metal. In all bronze pumps, the parts in direct contact with the process liquid are the pump manufacturer's standard bronze and the shaft is stainless steel or non-ferrous metal.

In steam pumps, the steam end is usually cast iron. Ductile iron or steel with a cast iron liner is used with high temperature steam. The steam piston rings are hammered iron rings. In non-lubricating service, two piece rings of bronze and hammered iron are used.

Reciprocating pumps are available in special alloys, stone wear and hard rubber for corrosive liquid applications. Usually special alloy cylinder liners are used on corrosive liquids. Brittle liners should not be under mechanical compression in the pump cylinder.

Cast iron is used up to 290°C (550°F). Cast steel, nodular iron, nickel iron or nickel steel are used at higher temperatures.

Viscosity and Speed

Viscosity is a fluid property defined by the equation:

$$R = \mu \, dv/dy$$

where R = shear stress (gm/cm sec^2), dv/dy = shear rate (sec^{-1}), μ = viscosity (poise).

---

*DuPont Trademark

With most fluids, the shear stress is proportioned to the shear rate. These fluids are called Newtonian fluids and examples are water, hydrocarbons and syrups.

Fluids where the shear stress and shear rate are not proportional are called non-Newtonian fluids. In these fluids, an apparent viscosity is measured at process conditions and is used to specify the pump. Apparent viscosity is defined by the equation:

$$\mu_o = \frac{R}{dv/dy}$$

When specifying a pump, the vendor should be given a curve showing viscosity response with changes in shear rate and shear stress for non-Newtonian fluids.

One type of non-Newtonian fluid is a dilatent liquid where an increase in shear rate causes an increase in apparent viscosity. Examples of dilatent fluids are slurries, clay and candy products. Another type of fluid is rheopectic fluids where the apparent viscosity increases with time for a constant shear rate.

A third type of non-Newtonian fluid is plastic fluids, which exhibit Newtonian behavior only above some minimum shear stress. Below that stress, they are solid. A fourth type is pseudo-plastic fluids where an increase in shear rate causes a decrease in apparent viscosity. Both plastic and pseudo-plastic behavior is independent of time. That is, when the shear stress stops, the fluid returns to its original viscosity. Examples of plastic and pseudo-plastic fluids are gels, latex paints, lotions and shortening. A fifth type is thixotropic where the apparent viscosity decreases as the shear rate increases and the liquid slowly returns to it original viscosity at zero shear. In addition, the apparent viscosity will decrease with time at constant shear rate so that the liquid will not return to its original viscosity with time at zero shear, but will remain at a lower value. Examples are greases, soaps, tars and peanut butter.

As a general rule, higher viscosities require lower pump speeds so that the liquid can fill the internal pump voids. The lower speed requires the specification of a larger pump to get the same capacity available from smaller pumps at lower speeds. Specifying a larger pump for high viscosity liquids is economical due to the lower maintenance costs and longer pump life of low speed pumps.

Lower pump speeds should also be used with abrasive and corrosive liquids. In both cases, high speeds will cause excessive wear.

An alternative to large, low speed pumps for high viscosity fluids is to increase the fluid temperature to decrease its viscosity. Steam jacketed pumps and jacketed piping can be used to keep a hot fluid warm.

When specifying a plunger pump, the speed at which liquid separates from the plunger should be calculated. This calculation can be done by the vendor once a particular pump is selected.

Slip

Slip is the amount of leakage between the discharge and suction side of a pump through small clearances in the pump. In reciprocating pumps, slip is due to leakage around check valves, stuffing box losses and fluid compressibility. It is expressed as a percentage of pump displacement and is usually less than five percent and in some pumps less than one percent. Entrained gas will compress in pumps and appear to be slip, but is not.

In rotary pumps, slip is primarily due to losses through small clearances. A relation defining slip through a clearance is: (30)

$$Q_s \propto \frac{\Delta P \; b \; h^3}{\mu \ell}$$

where Q_s = flow through the clearance, ΔP = the pump's differential pressure b = clearance width, h = clearance height, ℓ = length of the fluid path, and μ = viscosity. As the relation shows, slip is proportional to differential pressure and inversely proportional to viscosity. At viscosities above 1000 cp, slip reduces to zero in most rotary pumps.

Slip is proportional to clearance to the third power. Clearance varies with every pump depending on manufacturing tolerances and alignment. For this reason, each pump must be tested to determine the amount of slip present. Wear increases clearances and so increases the slip. Periodic checks of pump performance should be made and compared to manufacturer specifications for slip to estimate the pump wear rate. Slip is inversely proportional to the length of the fluid path through the clearance. This fact is applied in the design of lobe pumps with lobes that have long curved ends to reduce slip between the casing wall and the lobe.

Slip is not a function of speed in rotary pumps. This fact is used with low viscosity fluids, which are pumped at high speeds in small pumps to minimize slip losses. However, there is a speed at which the liquid will flash and flow will stop, especially in systems with a low NPSH available.

Pressure Head

The purpose of a pump is to overcome the pressure drop due to static head, system head and the suction and discharge line pressure drops. The system head and line drops should be calculated for each flow rate and at each process condition handled by the pump. In particular, head calculations should be made for each viscosity change since an increase in viscosity increases pipe friction head.

Friction head is calculated at peak flow rate which is 3.2 times the mean flow rate in a simplex single-acting reciprocating pump, 1.6 times mean flow in a duplex single-acting pump and 1.1 times mean flow in a triplex or higher single-acting pump (31). Friction factor charts and head calculation examples can be found in many standard reference texts (32).

Acceleration head, a part of pipe line pressure drop, is the pressure loss due to the acceleration of fluid mass in a pipe line. Acceleration head is usually small in rotary pumps, and so is ignored. However, reciprocating pumps with highly pulsed flow have high acceleration heads. Maximum acceleration head occurs at the point of maximum fluid acceleration, which is when the fluid flow rate is low and friction head losses are minimal. As a result, the maximum system pressure loss may occur with maximum acceleration (minimum flow) instead of maximum flow. Both cases should be checked.

Acceleration head can be calculated from the equation: (33)

$$h_a = \frac{L \, V \, n \, C}{K \, g_c}$$

where h_a = acceleration head in feet of liquid, L = length of suction line in feet, V = velocity in the suction line in feet per second, n = pump speed in rpm, C = pump constant, K = fluid factor and g_c = Newton's-law conversion factor. Values of C for different pump types are:

duplex single acting	=	0.200
duplex, double acting	=	0.115
all triplex	=	0.066
all quaduplex	=	0.08
all quintuplex	=	0.04
all sextuplex	=	0.055
all septuplex	=	0.028
all nonuplex	=	0.022

The above values may not hold for unusual connecting rod length to crank radius ratios. K = 1.4 for water, 2.5 for hot oil, and 1.0 for liquid with entrained gas. U.S. Hattiangadi gives the derivation of the acceleration head equation (34).

Once the system head losses are calculated, the pump vendor should be given the minimum differential pressure and maximum discharge pressure. Minimum differential pressure equals minimum discharge pressure minus maximum suction pressure. It is used to select a back pressure valve, if needed. Maximum discharge pressure is used to select a pump capable of handling that pressure. Maximum discharge pressure equals the process pressure plus the static head plus acceleration head plus the friction head loss plus 10% of the total. If there is no pump available that can pump against the maximum discharge pressure calculated, either select a lower speed or increase the discharge pipe diameter.

NPSH

Net positive suction head (NPSH) is the pressure energy needed to fill the pump inlet with liquid. The NPSH available at the pump inlet is calculated with the equation:

$$NPSH = Z_s + \left(144 \ \frac{(P_s - P_{vp})}{\rho} \right) - h_{fs}$$

where Z_s = static pressure head in feet of liquid at the pump suction. Z_s is negative in suction lift cases. P_s = absolute pressure at the suction pipe entrance in pounds per square inch, P_{vp} = liquid vapor pressure at the process temperature in psia, ρ = liquid density in pounds per cubic feet and h_{fs} = suction line head loss, including acceleration head, in feet of liquid. For rotary pumps, NPSH available is calculated at peak flow for the maximum line head loss. With reciprocating pumps, both peak flow and minimum flow (maximum acceleration head) conditions are used to calculate NPSH available, and the smaller of the two NPSH values is used in selecting a pump.

Every pump has a NPSH required, depending on the weight and spring load of the suction check valves and the pump speed. The pump manufacturer's literature gives the NPSH required for a pump at any given pump speed and fluid viscosity.

For plunger pumps, the NPSH required at the start of the stroke is calculated with the equation (35):

$$NPSH = \frac{\ell \ L \ D^2 \ N^2}{5.19 \times 10^4 \ D_i^2}$$

where ℓ = suction pipe length in feet, L = pump stroke length in inches, D = plunger diameter in inches, N = speed in strokes per minute and D_i = inside diameter of the suction piping in inches. The NPSH for the remaining cycle is calculated at the maximum flow. When specifying a plunger pump, the NPSH available should be about 20 to 35 kilopascals (3 to 5 psia) greater than the NPSH required, to avoid air leakage through the packing.

The NPSH available must exceed the NPSH required to avoid cavitation. Cavitation is due to vapor flashing at the pump suction and causes excessive wear and pitting in pumps.

Horsepower and Efficiency

The total power input to a pump is the sum of the driver and transmission losses, pump losses and the water horsepower. The water horsepower is the power that goes into the fluid and is calculated as follows (36):

$$WHP = \frac{Q\ \Delta P}{1714}$$

where Q = actual flow in gallons per minute and ΔP = differential pressure in psi across the pump. The power losses due to timing gears, bearings, packing and wear between the pump casing and rotor are mechanical losses and are exprssed as mechanical efficiency (E). It is defined as:

$$E = \left(\frac{WHP}{BHP}\right)100$$

where BHP = brake horse power, the actual power delivered by the driver. Mechanical efficiency is usually 90—95% at full load, depending on the pump, gears and fluid viscosity. Efficiency is inversely proportional to viscosity because the brake horsepower increases with increasing viscosity. Brake horsepower is the sum of water horsepower and viscous force losses (mechanical losses). The relationship between viscous force losses and viscosity is approximately

$$\frac{\text{mechanical losses}}{\text{mechanical losses}_1} = \left(\frac{\mu}{\mu_1}\right)^n$$

where the losses are measured at two different viscosities and a value for n calculated (37). Mechanical loss versus viscosity data is available from pump vendor literature. Mechanical losses with high viscosity fluids can be reduced by larger pump clearances.

The brake horsepower should be specified at the maximum fluid viscosity which usually occurs at start-up. In cases where the viscosity varies widely between start-up and normal operation, a two speed motor should be considered.

Mechanical loss versus viscosity data is available from pump vendor literature. Mechanical losses with high viscosity fluids can be reduced by increasing the fluid temperature.

Reciprocating pumps are commonly rated by volumetric efficiency. Volumetric efficiency is the percentage of the actual displacement to the theoretical displacement. In a single, double-acting piston pump, the theoretical displacement is:

$$D_T = \frac{12\ A\ S}{231}$$

where D_T = theoretical displacement in gpm, A = piston area in inches^2, and S = piston speed in feet per minute. For a duplex pump, multiply by 2. The actual displacement can be obtained from vendor literature. The difference between theoretical and actual displacement is slip, which in this case includes the piston rod cross sectional area. Volumetric efficiency increases with speed because slip is constant with changes in speed while displacement increases. Entrained air reduces volumetric efficiency and should be minimized.

Torque

In rotary pumps, the required shaft torque should be calculated in low speed-high horsepower applications, such as with high-viscosity liquids. The torque can be computed from the following equation (38):

$$BHP = 1.903 \times 10^{-4}\ N\ T_P$$

where N = speed in revolutions per minute and T_P = torque in foot-pounds. The required torque must be less than the rated torque of the pump to avoid damaging the pump.

Start-up and Troubleshooting

During the installation of a pump, temporary screens should be put in the suction piping to protect the pump. Also, pressure-vacuum gauges should be mounted on the suction and discharge piping to monitor pump performance. Before starting up a pump, a check list should be made outlining the start-up procedure.

In general, the first item on a start-up list is to recheck the pump and drive train alignment and the piping supports and alignment. The suction piping should be tested to see that it is air tight in cases when a pump has a suction lift. Otherwise, a water test is sufficient. Secondly, rotate the pump shaft by hand to check for internal blockage. Lock out the motor starter before turning the pump shaft. Third, start-stop the motor to check the direction of rotation. Fourth, check the relief valve setting and action. With reciprocating pumps, the relief valve is often used as a by-pass to unload the pump during start-up, unless a by-pass line exists. Fifth, prime the pump, if necessary. A rotary pump should be at least wetted with liquid before starting to reduce wear and improve suction lift. Sixth, open all suction and discharge piping valves. Seventh, lubricate the packing, seals and bearings and turn on any flush liquids. Eight, start the pump. The pump should deliver fluid in about a minute.

If it does not, turn it off and determine the problem. In self-priming applications, it may take several minutes to lift liquid through the suction lines to the pump depending on the suction line length.

Some of the more common pumping problems and their causes are described here.

1. Pump shaft rotating but no liquid discharged:

Pump not primed; insufficient available NPSH; suction-line strainer clogged; end of suction line not in liquid; relief valve set lower than minimum required discharge pressure or jammed open by foreign material; pump rotating in wrong direction; suction or discharge valves closed; by-pass valve open; insufficient liquid supply.

2. No liquid discharge-steam pump:

Low steam pressure; water in steam cylinders or steam line.

3. Pump delivers below rated capacity — Rotary Pump:

Air leaks in the suction line or through the seals; insufficient NPSH; suction line strainer partially clogged or has insufficient open area; excessively worn pump intervals leading to excess slip; wrong pressure on relief valve; relief valve jammed open; speed too low; suction or discharge valves partially closed; by-pass open; liquid viscosity different than specified; liquid vaporizes in suction line; excessive system pressure; obstruction in the discharge line; liquid temperature and vapor pressure higher than expected; wrong sized rotor; pump not filling; vortex in supply tank; liquid degassing at suction.

4. Pump delivers below rated capacity — Power Pump:

All the causes listed for rotary pumps; one or more cylinders running dry; pump check valve stuck open.

5. Pump delivers below rated capacity — Steam Pump:

Low steam pressure; insufficient lost motion in valve gear; excessive steam cushion at ends of cylinder; insufficient steam cushion; excessive lost motion; little or no steam exhaust pressure; packing too tight.

6. Pump will not turn:

Motor not running; keys sheared or missing; drive belts, power transmission components slipping or broken; pump shaft sheared.

7. Pump loses prime:

Liquid level falls below the suction line intake; air leak in pump seal; air leak in suction line; liquid vaporizes or degasses in suction line; liquid drains or syphons from system during off-periods; worn rotors; insufficient NPSH; in a vacuum inlet system at the initial start-up, atmospheric "blow back" prevents pump from developing enough differential pressure to start flow.

8. Excessive power consumption:

Speed too high; shaft packing too tight; liquid more viscous than specified; misalignment; obstruction in discharge line raises operating pressure; discharge line too small; discharge valve partially closed; lube oil too viscous; pipe strain on casing; excessive discharge pressure; liquid sets up in line and pump during shutdown; poor lubrication.

9. Excessive heat:

Same causes as excessive power consumption.

10. Noisy operation — Rotary Pumps:

Cavitation; misalignment; foreign object in pump; bent rotor shaft; relief valve chattering; running dry; air leakage in suction piping or stuffing box; insufficient liquid supply; internal elements binding or loose; pipe strain on casing; relief valve chattering; obstruction in discharge line; dissolved gas degassing at pump suction; worn bearings or timing gears; too high speed (high speed pumps are not quiet — check only erratic sounds).

11. Noisy operation — Steam Pumps:

Piston loose on rod; steam piston hitting head; liquid valves hanging up.

12. Noisy operation — Power Pumps:

Piping too small and/or too long; worn valves or seats; valves stuck open.

13. Rapid wear:

Pipe strain on pump; grit or abrasive material in liquid; pump running dry; corrosion; misalignment; rotating elements binding; pipe strain on casing; excessive system pressure; speeds higher than rated; poor lubrication of bearings and gears.

REFERENCES

1. Berk, W.L., Water and Sewage Works — Reference No. 1976, April 30, 1976, p. R-22.
2. Hicks, T.G., and Edwards, T.W., Pump Aplication Engineering, McGraw-Hill Inc., 1971, p. 322.
3. Karassik, I.J., Krutzsch, W.C., Fraser, W.H., and Messina, J.P., Pump Handbook, McGraw-Hill Inc., 1976, p. 3—53.
4. McKelvey, J.M., Maire, U., and Haupt, F., Chem. Engr., *83*, Sept. 27, 1976, p. 101.
5. Boulden, L., Machine Design, *45*, June 28, 1973, p. 99—100.
6. Holland, F.A., and Chapman, F.S., Chem. Eng., *73*, Feb. 14, 1966, p. 132.
7. Hicks, T.G., and Edwards, T.W., ibid, p. 241.
8. Ibid, p. 56.
9. Karassik, I.J., et al, ibid, p. 3—36.
10. Ibid, p. 3—12.
11. Ibid, p. 3—7.
12. Hicks, T.G., and Edwards, T.W., ibid, p. 47.
13. Karassik, I.J., et al, ibid, p. 3—14.
14. Ibid, p. 3—16.
15. Proportioneers, Division of General Signal Corp., West Warwick, R.I. Bulletin 1000.22, p. 3.
16. Karassik, I.J., et al, ibid, p. 3—23.
17. Worthington Pumps, Rotary and Centrifugal Pump Theory and Design, Worthington Pump Corp., 1971, p. 50.
18. Karassik, I.J., et al, ibid, p. 3—23.
19. Proportioneers, ibid, p. 1.
20. Karassik, I.J., et al, ibid, p. 3—23.
21. Goulds Pumps Inc., GPM Goulds Pump Manual, Goulds Pump Inc., New York, 1976, p. 329.
22. Ibid, p. 330.
23. Holland, F.A., and Chapman, F.S., ibid, p. 139.
24. Worthington Pumps, ibid, p. 49.

25. Holland, F.A., and Chapman, F.S., ibid, p. 141.
26. Rost, M., and Visich, E.T., Chem. Engr./Deskbook Issue, *76*, April 14, 1969, p. 56.
27. Thurlow, C., Chem. Engr., *72*, June 7, 1965, p. 213.
28. Hydraulic Institute, Hydraulic Institute Standards for Centrifugal, Rotary and Reciprocating Pumps, 13th ed., 1975, p. 124.
29. Boulden, L., ibid, p. 100.
30. Waukesha Foundry, Waukesha Pump Engineering Manual, Abex Corp., Wisconsin, 1975, p. 26.
31. Hattiangadi, U.S., Chem. Eng., *77*, Feb. 23, 1970, p. 108.
32. Perry, R.H., Chilton, C.H., and Kirkpatrick, S.D., Chemical Engineer's Handbook, 4th ed., McGraw-Hill Inc., 1963, p. 5—20.
33. Hydraulic Institute, ibid, p. 226.
34. Hattiangadi, U.S., ibid, p. 107.
35. Holland, F.A., and Chapman, F.S., ibid, p. 150.
36. Waukesha Foundry, ibid, p. 33.
37. Zalis, A.A., Pet. Refinery, *40*, Sept. 1961.
38. Karassik, I.J., et al, ibid, p. 3—95.

CHAPTER 4

SEALS AND PACKINGS

K. S. PANESAR
Houston, Texas

Anything and everything that moves or does any work eventually wears out and must be either replaced or repaired. How long a piece of equipment lasts will depend upon many factors, like speed, environment (pressure, temperature, and chemical properties of the fluid being handled, etc.) and maintenance. With proper maintenance you not only prevent premature equipment failure, but you also prolong the life of the machinery. Almost all machines need maintenance, and mechanical seals are one of the most delicate pieces of equipment that require such special care. In order to cut down maintenance, the seals must be properly selected for each job. If you asked the maintenance department of any petrochemical plant or refinery to name one item that gives the most problem, the answer in the majority of the cases would be "SEALS." The selection of the seals, therefore, is very important.

Packings have been used as sealing devices for many years and in some special cases they are still used. However, mechanical seals have taken over most application today. Let us discuss the advantages and disadvantages of both mechanical seals and packings.

PACKINGS

Packings form a seal between the throat of a stuffing box and its gland. Normally six rings of packing are used and in the middle a lantern ring is used for flushing, cooling and lubricating the packing. Packings can be round, square, or cone shaped and made from cloth, jute, Teflon, carbon, and even metals.

Advantages

1. Packings are simple and easy to replace.
2. Packings are very inexpensive compared to mechanical seals.
3. If the rings get a little worn and start to leak at more than usual rate, the gland bolts are tightened a little to reduce leakage.
4. No specially trained mechanics are required to install or replace packings.

Disadvantages

1. Packings leak too much as compared with the seals. This is dangerous if the pumping product is toxic or explosive.
2. Packings can burn at high speeds.
3. If tightened too much, packings can damage and even break the shaft.
4. Packings usually require more maintenance than seals.

Basic Construction

Even though mechanical seals are taking over the majority of the field, packings are still used in some pumping applications. A simple application being in water service, where leakage to atmosphere is tolerated. Packings will probably never be replaced by mechanical seals, in the valve stem and reciprocating pumps and compressors application. Packings can be of square, rectangular or circular cross-section, but square type is the most popular. These packings are made from various materials like animal, vegetable, metals and synthetic fibers. But the modern trend is towards metals and synthetic materials like teflon and graphite. Metal packings are made from lead, babbitt, copper or aluminum and are used where strength and heat-resistant properties are desired. Asbestos, graphite and Teflon are popular over a wide range of applications. In reciprocating compressors, especially for non-lube application, teflon is the most popular choice.

Types of Packings

There are basically three types of packagings: (1) compression type, (2) automatic and (3) floating type.

Compression Type — This is the most commonly used in pumps and the force is applied by tightening the gland bolts. Packings must be lubricated in this case and if somehow lubricant is lost, the packing will become hard and damaged and should be replaced.

Automatic Type — In this type usually pressure creates the sealing force.

Floating Type — The packings for reciprocating compressors are usually segmented and are held in place by springs. This type is called floating type.

Packing Selection

Packing should be compatible with the fluid being pumped. Teflon packings have a wide use in the industry because of its inertness to many chemicals. Similarly, asbestos finds more use in high temperature applications. Packings for pumps in dirty or slurry service should be flushed with external clean fluid. This is accomplished with a lantern ring which is made of brass, stainless steel or teflon depending on pumpage. Usually, five to six packing rings are used and the lantern ring is inserted in the middle as shown in Figure 4.1. The clean external fluid is used anywhere from 15 to 25 psi above the stuffing box pressure. This keeps the dirty fluid from entering under the packings and thus preventing the shaft or shaft sleeve from getting damaged. The clean liquid lubrictes the packing and also absorbs some of the frictional heat. Also clean liquid leaks out to the atmosphere instead of dirty or toxic fluid. One of the most common guides used in the selection of the packings is the "PV" factor, where "P" is the pressure (in PSIG) in the stuffing box and "V" is the shaft velocity (in feet per minute). It is a measure of heat input in a packing or degree of difficulty of packing. The higher the number, the more difficult (careful) the selection. The Figure 4.2 PV graph shows the different types of packing for different PV values.

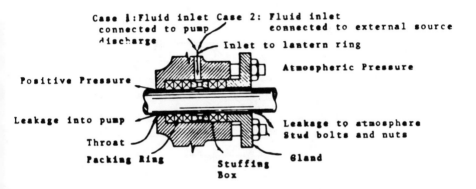

Figure 4.1. A typical stuffing box showing lantern-ring. Case 1 is commonly used when the fluid is clean and unharmful to the personnel. Case 2 is used when the fluids are dirty or toxic. In case of highly abrasive fluids or slurries, the lantern ring is usually moved to the left (bottom of the stuffing box) and clean external fluid is used so that the abrasive material does not get under the packings at all.

Another factor to consider is the pH-value for packings compatibility with the fluid to be pumped. Some manufacturers are now offering packings made from materials that can handle pH values ranging from 0−12. One such packing is made of equal mixture of Teflon and graphite.

Maintenance

Packings are more prone to maintenance than the mechanical seals. After a while they wear out and fluid starts to leak more than usual and this is taken care of by just tightening the gland bolts. But, after some time the packing needs to be replaced[1]. Shaft sleeves should be used whenever packing is employed. This saves the shaft from the wear and tear by the packing. instead, the shaft-

---

[1]Do not add another ring. Just replace the worn out ring with a new ring.

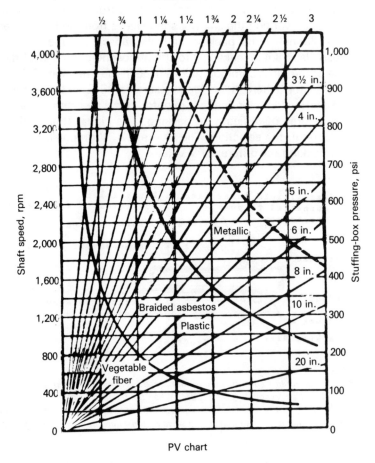

Figure 4.2.

sleeve gets worn out and is more economical to replace than the shaft. The sleeve under the packing should have good and smooth finish and may be hardened to prolong life.

MECHANICAL SEALS

Advantages

1. Very little or no leakage, if installed properly.
2. Less of product dilution or contamination when using external flush.
3. EPA(Environmental Protection Agency) and OSHA (Occupational Safe-

ty and Health Administration) have imposed stringent rules for the leakage of dangerous and toxic fluids into the environment. With double seals the leakage of poisonous fluids can be almost eliminated.

Disadvantages

1. Seals are more expensive than packings.
2. Skilled labor is needed to install and repair a mechanical seal.
3. Handled improperly, seals can cause expensive shutdowns.

Basic Construction

Basically a mechanical seal forms a dynamic seal between two flat surfaces perpendicular to the rotating shaft. One piece or face stays stationary while the other face runs or rotates with the shaft. (See Figure 4.3). This is called the primary seal. There is a 'secondary seal' between the shaft and the seal head. This is usually in the form of an "O" ring, but can be in other shapes like wedge, V-shape or U-shape. The most commonly used materials for the secondary seal are elastomers like Viton, Teflon, etc. There is a third major

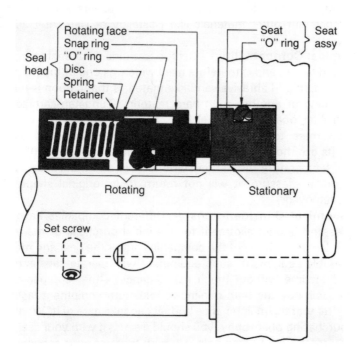

Figure 4.3. Parts of a mechanical seal. (Courtesy John Crane Packing Co.)

component called the spring which keeps the faces (primary seal) clsoed most of the time. The hydraulic forces (fluid pressure in the stuffing box) also help the spring force in keeping the faces together. The spring is held in a sub-assembly attached to the shaft or shaft sleeve.

Selection Parameters

The selection of materials for the seals is usually based on three factors:
a. Mechanical properties
b. Chemical compatibility, (corrosion resistance) and
c. Economical considerations.

a. Mechanical Properties — The two faces form the primary seal by rubbing together. They generate considerable heat. The materials selected should be able to withstand the heat without losing their strength. Two hard materials rubbing together will generate even more heat than do soft and hard. One of the face materials chosen is soft, and carbon has been found to be the best candidate for this service. The second face material should be hard and should be applicable almost universally. Tungsten-Carbide has been found to meet most of the requirements for a variety of fluids, pressures and temperatures. The industry has standardized on stainless-steel springs and their assembly, but other materials like Hastalloy or Carpenter 20 are also available.

b. Chemical Compatibility — Carbon is an inert material to many fluids being pumped today, and is therefore used widely as one of the faces. Only in cases of very hard and abrasive sand-like materials that carbon is replaced by tungsten-carbide or another such hard material. Tungsten-carbide has two grades available; one is the nickel base and the other is the cobalt base. Among elastomers, teflon is one material that is practically inert to almost all chemicals. Its use, however, is limited to temperatures below 500°F, because around this temperature the Teflon starts to flow and if a Teflon "O" ring is deformed under pressure, it will not return to its original shape after the pressure is removed.

c. Economical Considerations — Different companies have different philosophies and budget allotment for the initial purchase of the equipment and its maintenance. Some of the companies prefer the best and most expensive and reliable equipment; such equipment will operate satisfactorily for a long period of time without much maintenance. Other companies, on the other hand, will buy the bare minimum; minimum equipment lasts about a year, and after that the maintenance budget can take care of it. Whatever your needs or purchasing philosophy, you should discuss it with your seal manufacturer's representative who can help you with most of your applications.

MECHANICAL SEAL CONFIGURATIONS AND DESIGN

Basically seals can be divided into two classes; the Pusher type and the

Non-pusher type (See Figure 4.4a and 4.4b). In the pusher type, there is a secondary sealing member, i.e. the elastomer, which *pushes* along the shaft as the faces wear out. In the non-pusher type, however, there is no secondary sealing device (elastomer) and there is no relative movement between the shaft and the seal. The axial movement is taken up by the bellows to keep the faces closed and maintain the primary seal. Next you can have seals that are called Balanced or Unbalanced by the nature of their design. Rubber or metal bellows seals are balanced by design. Seals are also classified by their arrangement: Single, inside, outside, double, etc. Table 4.1 shows the different designs and arrangements used for sealing various types of fluids. Let us discuss the main features of each step by step.

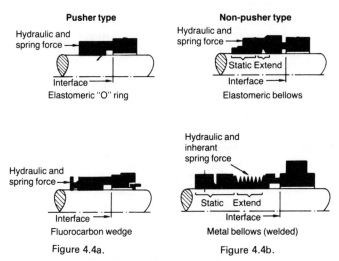

Figure 4.4a.   Figure 4.4b.

Figure 4. Pusher and non-pusher seal types. (Courtesy John Crane Packing Co.)

Table 4.1. Mechanical Seals Configurations & Designs

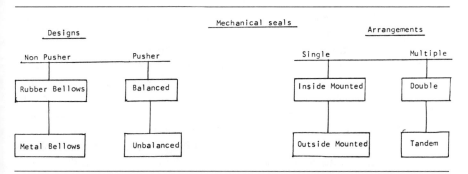

Single Seals

Single seals are used in handling relatively clean fluids, which also lubricate the faces. If the fluid is a little bit dirty and the minute amounts of solid particles can be cleaned by using cyclone separators, then the pumpage is used for flushing. If the fluid is extremely dirty and cannot be used for self flush, then an external slush (API Seal Flush Plan 32) is employed. Dirty fluid can damage the elastomer, clog up the springs, and in some cases even damage the seal faces, more particularly the carbon faces.

a. **Single Inside Seals** — In the chemical and petrochemical industry, where flammable and hazardous materials are normally pumped, the seals are mounted on the shaft inside the stuffing box. The pumping fluid is thus contained in the pump housing.

b. **Single Outside Seals** — In the food industry, especially where health and cleanliness are of utmost importance, seals are mounted outside the stuffing box or the pump housing. This way they can be cleaned more easily than an inside seal.

c. **Single Unbalanced Seals** — API-610, for centrifugal pumps for general refinery service, paragraph L, limits the use of unbalanced seals as shown in Table 4.2. below:

Table 4.2. Limits for Unbalanced Seals

Seal I.D. Inches	Shaft Speed Rpm	Sealing Press Psig
1/2 to 2	up to 1800	100
	1801 to 3600	50
Over 2 to 4	up to 1800	50
	1801 to 3600	25

Unbalanced seals are less expensive but also have limited use. For instance, they can be used up to a maximum of about 350 psig, after which balanced seals should be used. See Figure 4.5a for the unbalanced design.

d. **Single Balanced Seals** — Balance is achieved by cutting a step in the shaft or the sleeve over the shaft, and thus the rotating face is lowered. Hydraulic pressure acts against only part of the face area. See Figure 4.5b for this balanced design. A complete hydraulic balance is achieved by further lowering the face to a point where hydraulic pressure does not exert any force on the sealing area. Only the spring exerts force against the sealing area to keep the seal faces closed. See Figure 4.5c for the complete balanced design. This method of reducing load at the seal faces helps to prolong seal life. Balanced seals are used in high speed and high pressure applications.

e. **Double Seals** — These seals are used whenever you are pumping fluids that are dangerous and carry health hazards. With double seals a buffer zone is

Unbalanced

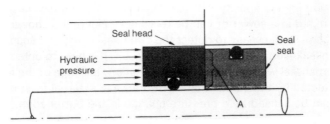

An unbalanced seat-head combination. The total hydraulic pressure acts against the rear of the head and on the sealing area A.

Figure 4.5a. Unbalanced

Partial balance

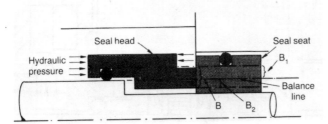

Partial balance achieved when head sealing face is lowered by means of a step cut in the sleeve. Hydraulic pressure acts against a portion of the total face area B. The factor of "per cent of balance" is created by distributing the area of B_1 and B_2 above and below the balance line respectively.

Figure 4.5b Partial balance.

Complete balance

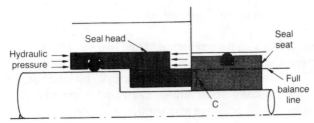

Completely balanced combination. The head seal face has been lowered to a point where the hydraulic pressure does not exert any load against the sealing area C. Only spring pressure acts against the sealing faces.

Figure 4.5c. Complete balance

Figure 4.5. Balanced/unbalanced seal designs. (Courtesy John Crane Packing Co.)

created in the stuffing box, and a clean compatible external flush is used at about 15 to 25 psi above the stuffing box pressure (normally equal to the suction pressure plus a little more). Thus the pumpage never leaks to the atmosphere. A nominal flow of about 1 to 4 gpm is adequate for this external flush and you should specify API seal flush plan 33. The discharge from this can be dumped in a sewer or can be reclaimed. Figure 4.6 shows the double seal arrangement. It is easy to detect the outer seal's failure, because the fluid will start to leak along the shaft and can be seen. It is hard to detect the failure of the internal seal without some instrumentation. Leakage of the internal seal can be detected by the contamination of the clean external flush in the buffer zone, or can be sensed by a pressure change in the buffer zone.

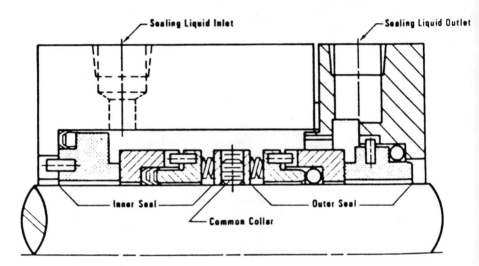

Figure 4.6. Showing the double seal design. (Courtesy Durametallic Corp.)

f. Tandem Seals — This arrangement has two seals facing the same direction. The internal seal gets its flush from the pump discharge and the outer seal gets its lubrication from an external source. This arrangement is usually used in the high pressure applications. Each seal is capable of withstanding the full pressure. These seals are also used in environments where safety is of paramount importance. Figure 4.7 shows the tandem arrangement. The seal failure can be detected in a similar way as mentioned above in the case of double seals.

Seal Selection

Seal selection is not as easy as it appears. The person responsible for the seal selection must know and inform the manufacturer of the seal about the properties of the fluid being pumped. He should specify the pressure, temperature, vapor-pressure, viscosity, corrosive properties, if any, solid particles or abrasives in the pumpage, and so forth. Of all the above, the

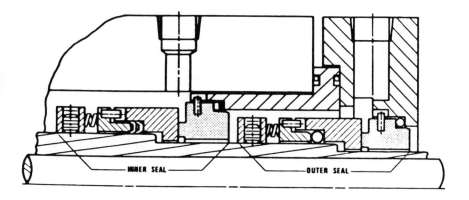

Figure 4.7. Showing the double tandem seal arrangement. (Courtesy Durametallic Corp.)

temperature and the chemical properties are the most important factors.

Temperature

This is a very critical parameter and should be given proper attention, especially when using the pusher type seals, which employ elastomers. Most elastomers have a limit under which they operate satisfactorily; above that limit they decompose or lose their sealing effectiveness. The most common elastomers used in the pump industry are: Fluorocarbons (Teflon), Buna-N, Neoprene, and Viton, API-610, appendix D, under general notes for mechanical seals, lists the temperature limits for these elastomers. The limits are shown in Table 4.3.

Table 4.3.

Gasket Material	Min.Temp. °F	Max. Temp. °F
Fluorocarbon	− 150	+ 500
Buna-N	− 40	+ 250
Neoprene	Zero	+ 200
Viton	Zero	+ 400
*Metal Bellows	No limit	No limit

*Note: Metal bellows, especially for high temperature have no elastomers, and have no temperature limitations.

It is a good practice to switch to metal bellows above 450 °F. The manufacturers are always working to develop new materials, and a few years ago they developed a new elastomer called "Kalrez." The manufacturers claim "Kalrez" can stand up to 600 °F, but how long it will last at that temperature remains questionable. Kalrez, however, has one advantage over Teflon; it does not flow like Teflon and it has memory, i.e. this material will come back to its original shape after the deforming force is removed.

The fluid temperature in the stuffing box must be carefully evaluated because the two faces rubbing together also generate heat. If no cooling is provided, the fluid can flash and lose its lubricity. The seal faces then run dry and can be easily damaged. In such cases the stuffing box jackets must be cooled and, if necessary, the pumpage should also be cooled. John Crane Packing Company recommends that you cool the fluid 50°F below the *Atmospheric Boiling Point* and use about one gallon per minute of cooling medium for every inch of diameter of the seal. High temperature metal bellows seals have been used in the refineries without direct cooling.

These seals were dead-ended in the stuffing box, and the fluid temperature was 700°F plus. There was no product cooling except the stuffing-box jacket, which was water cooled, and a steam quench on the outside of the seal as according to seal flush plan 62, API-610. The purpose of the steam quench was two-fold: First, it cleaned any coking particles which could clog up the passages underneath the faces, and the second, since low pressure steam was used, its temperature was well below the pumpage and thus provided some cooling for the seal. Most plants have 50-Psig steam available. A stainless steel needle valve is needed to throttle the amount of steam needed for quenching down to 2 to 5 Psig. Care should be taken not to use too much steam, because the excess steam can condense along the shaft and can contaminate the bearing oil, thereby damaging the bearings.

Chemical Properties

Even though Teflon causes shaft fretting (or sleeve fretting), its use is widely spread because of its special property of being inert to most chemicals. Viton is the next most commonly used material in the petrochemical industry. While Viton is satisfactory for most hydrocarbons up to 400°F, it should be avoided in the amine solutions containing carbon dioxide, hydrogen-sulfide, etc. Similarly for hot water, condensates, and boiler-feed-water services, tungsten carbide with nickel is preferred over tungsten carbide with cobalt binder. For ammonia service, Viton should again be avoided; instead, Neoprene or ethylene-propylene (sometimes designated as E-P) are recommended. Manufacturers are always trying to develop new materials. In case you have a special application it is best to discuss the complete details with the manufacturers. Communication, verbal as well as written, is very important in selecting the right seal for the right job.

SEALING SPECIAL FLUIDS

In chemical and petrochemical industry, you come across a wide range of fluids that are easy to handle under one set of conditions, but are very difficult to handle under a different one. Let us discuss some of those fluids and their sealing.

1. Fluids That Crystallize

Sugar syrup, salt solutions and sulfur are some of the fluids that crystallize when exposed to the atmosphere and low temperatures. When pumping sugar syrup, for instance, keep the temperature in the stuffing box constant and use balanced single seal with external flush (API Plan 32) using compatible fluids, i.e. water, condensate or steam. In the case of sulfur, the pump housing and the stuffing box jackets should be steam jacketed (or steam traced) and steam quench should be used as the auxiliary seal flush (API Plan 62). For thermosensitive fluids use external flush (API Plan 32 or 33) at the same temperature as the pumping fluid.

2. Hot Products

Different materials have different coefficients of expansion and thus expand differently. The stationary ring, if only pressed fitted can fall out when pumping hot fluids, and the seal can start leaking. The stationary seat in such cases should be attached to the housing by a pin or some other positive means, and the following should be done:

a. Cool the pumpage by circulating water through the jackets and or use external flush. If self-flush employed (API Plan 11, etc.) the product can be cooled by using small heat exchanger, i.e. by using API Plan 21. For boiler feed water service, API-23 is usually preferred. Whenever water temperature exceeds 170°F, it is advisable to use a small heat-exchanger in the flushing line to cool down the water to about 160°F or less, before it reaches the seal faces. The quantity of cooling water required for use in the heat exchanger would depend upon the pumping temperature. For example, it would vary between 3 to 4 gpm for pumpage temperatures up to about 250°F. It should be noted that these quantities are for the case when the stuffing-box jackets also have cooling water circulating through them, otherwise the cooling water requirement would be a little higher.

b. Switch to metal bellow seals without any elastomers; as elastomers are not very good above 400°F. Asbestos is commonly used for high temperature sealing, but it is not a good material to be used as a secondary sealing member. Asbestos does not slide like the common elastomers and therefore can be damaged in pusher type seals. Asbestos is acceptable in seals where very little movement back and forth is experienced. Grafoil, on the other hand, is a satisfactory material for this application.

3. Fluids That Set-Up or Harden

Materials like asphalt, paint and latex, etc. tend to harden and set-up when cooled or exposed to air. To prevent them from setting-up or hardening, the stuffing box should be steam jacketed, and steam should be used per API Plan 62 as a quenching medium. If product contamination can be tolerated, then you could use external seal flush plan 32; otherwise you should use seal flush plan 33 for double seals.

4. Dirty Abrasive Products

For fluids that are dirty or contain highly abrasive materials like sand, double seals or single seal with external flush is highly desirable. Sand particles can get embedded in the carbon and rub against the tungsten-carbide face. This can make grooves in the tungsten carbide and the faces can start to leak more than normal. In such cases, you may use both stationary and the rotating faces of the same material, i.e. tungsten-carbide against tung-carbide. If the fluid contains only minute amounts of solids they may be removed by using cyclone separator (API Plan 31.) If the fluid happens to be hot, you could use API Plan 41. Some users do not like cyclone separators because they are additional maintenance items. They can also clog up and can cause problems.

5. Cryogenic Products

Products like ethane and ethylene are pumped at approximately $-150\,°F$ or so, and tend to ice up when exposed to the atmosphere. This can damage the seal and in some cases even the shaft. To prevent this from happening, auxiliary seal flush plan 52 should be used in conjunction with seal flush plan 11. Methanol is the most common fluid used for this external flush and only a small amount of injection is enough to prevent the pumpage from icing. You can have an alarm and shutdown in case of seal failure. You may specify auxiliary shaft-sealing by using at least two rings of packing between the seal and the end-plate (See Figure 4.8).

6. Dangerous/Toxic Fluids

The list of dangerous fluids (including acids) is too long to be included in this text. Only a general method of handling such fluids will be discussed here. In petrochemcial industries and refineries, H_2S is one of the common trace elements pumped along with amines. H_2S is lethal in half an hour in concentrations of 800 ppm to 1000 ppm. Similarly, hydrogen cyanide (HCN), is lethal in half an hour exposure to concentrations of 100 ppm to 200 ppm. The leakage of such elements into the atmosphere is prohibited by EPA and OSHA. To prevent leakage of such dangerous elements, double seals are

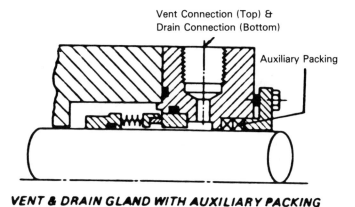

Vent Connection (Top) &
Drain Connection (Bottom)

Auxiliary Packing

VENT & DRAIN GLAND WITH AUXILIARY PACKING

Figure 4.8. Showing the primary seal and the auxiliary seal with two rings of packing. (Courtesy Sealol, Inc.).

highly recommended. Single seals with an external fluid flushes would be all right, if the product contamination is not objectionable. NOTE: For schematics of API flush plans mentioned above, refer to Figures 4.9 and 4.10.

SPECIAL PROBLEMS/TROUBLE-SHOOTING

1. Corrosion

If the carbon ring is corroded away or crumbled or blistered then the quality of the carbon is poor. To remedy the situation, switch to a new and better quality of carbon. If there is a hole in the carbon, it could be from the flush pointing at the carbon. In that case change the direction of the impingement of the flush.

2. Heat Check

"Heat Check" is the cracking of the rotating ring and it is caused by excessive heat generated due to lack of lubrication of the faces. The remedy is to provide cooling/lubrication. If this does not help, change the material of the rings. Stellite and ceramic rings are pretty susceptible to damage due to heat check. Tungsten-carbide is probably one of the best materials to resist this kind of damage.

3. Metal Bellows Fail

Sometimes metal bellows will fail or break at their convolutions. Unless it is fatigue failure, it could be due to the flush pointed at the convolutions. To correct the problem, just change the direction of the flush. Even though not very

167

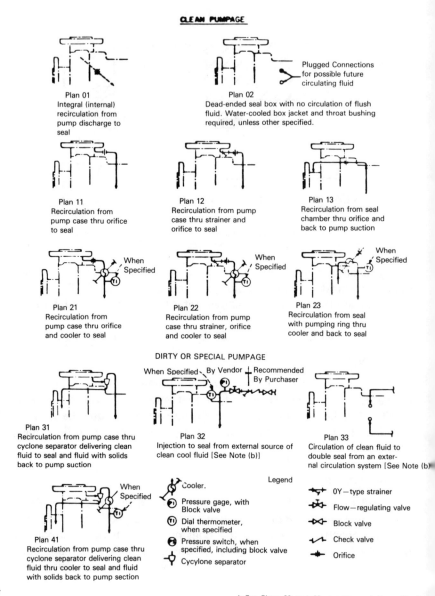

CLEAN PUMPAGE

Plan 01
Integral (internal) recirculation from pump discharge to seal

Plan 02
Dead-ended seal box with no circulation of flush fluid. Water-cooled box jacket and throat bushing required, unless other specified.

Plugged Connections for possible future circulating fluid

Plan 11
Recirculation from pump case thru orifice to seal

Plan 12
Recirculation from pump case thru strainer and orifice to seal

Plan 13
Recirculation from seal chamber thru orifice and back to pump suction

Plan 21
Recirculation from pump case thru orifice and cooler to seal

When Specified

Plan 22
Recirculation from pump case thru strainer, orifice and cooler to seal

When Specified

Plan 23
Recirculation from seal with pumping ring thru cooler and back to seal

When Specified

DIRTY OR SPECIAL PUMPAGE

Plan 31
Recirculation from pump case thru cyclone separator delivering clean fluid to seal and fluid with solids back to pump suction

Plan 32
Injection to seal from external source of clean cool fluid [See Note (b)]

When Specified   By Vendor   Recommended By Purchaser

Plan 33
Circulation of clean fluid to double seal from an external circulation system [See Note (b)]

Plan 41
Recirculation from pump case thru cyclone separator delivering clean fluid thru cooler to seal and fluid with solids back to pump section

When Specified

Legend

Cooler.

Pressure gage, with Block valve

Dial thermometer, when specified

Pressure switch, when specified, including block valve

Cycylone separator

0Y — type strainer

Flow — regulating valve

Block valve

Check valve

Orifice

Notes:

a. These plans represent commonly used systems. Other variations and systems are available, and should be specified in detail by purchaser or as mutually agreed between purchaser and vendor.

b. For Plans 32 and 33, purchaser shall specify the characteristics, and vendor shall specify the required gallon minute (gpm) and pounds per square inch gage (psig).

Figure 4.9. Piping for primary seals. (Courtesy API).

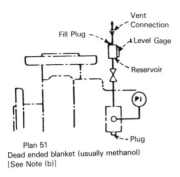

Plan 51
Dead ended blanket (usually methanol)
[See Note (b)]

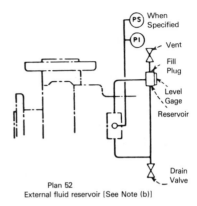

Plan 52
External fluid reservoir [See Note (b)]

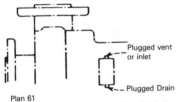

Plan 61
Tapped connections for purchaser's use. Note (b) shall apply when purchaser is to supply fluid (steam, gas, water, other) to auxiliary sealing device

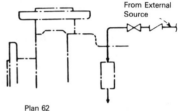

Plan 62
External fluid quench (steam, gas, water, other) [See Note (b)]

Notes:

a. These plans represent commonly used systems. Other variations and systems are available, and should be specified in detail by purchaser or as mutually agreed between purchaser and vendor.

b. For Plans 51, 52, 61, and 62, purchaser shall specify fluid characteristics when supplemental seal fluid is provided. Vendor shall specify the required gallons per minute (gpm) and pounds per square gage (psig) where these are factors, for example, when auxiliary seal is of the outside mechanical type.

Figure 4.10. Piping for throttle bushing or auxiliary seal device. (Courtesy API).

common, sometimes either the first or the last convolution may fail or break. The reason for this is that these convolutions are incomplete; they are chopped off to make the attachment to the seat and the face, and thus are mechanically the weakest areas.

4. Installation & Repair

As easy as it may look, the installation of the mechanical seal should be done very carefully, because the elastomer can be damaged very easily at the time of installing a new seal. The elastomer can hit the sharp edge of the shaft and can be easily damaged without being noticed. It is good practice to have shaft, sleeve or step in the shaft or sleeve's edge chamfered and smoothed out before installing a new seal.

5. Start-Up Problem

Sometimes due to hurry or oversight, the operators may start a pump dry, i.e. without fluid in the pump housing or in the stuffing box. This can burn up the seal. It is, therefore, advisable to make sure that the pump housing and the stuffing box are full of fluid. On very hot and very cold operations, it is again recommended that the pump be brought to the operating temperature gradually NOT instantly.

ACKNOWLEDGEMENTS

The writer is very thankful to Mr. Harry Tankus, President of the John Crane Company for his permission to use his diagrams and material.

Acknowledgments are also due to Mr. Bill Carpenter, Head of the OEM and Project Marketing, Durametallic Corporation, for permission to use their illustrations.

The writer is indebted to M/s Marty Doddel and Ray Stromoski of Sealol Inc. for their help and permission to use their diagram.

REFERENCES

1. API-610: Centrifugal Pumps for General Refinery Services.
2. *Dangerous Properties of Industrial Materials* by N. Irving Sax Van Nostrand Publishing Company.

CHAPTER 5

DRIVES FOR PROCESS EQUIPMENT

W. W. Willoughby
Senior Engineer
Dravo Engineers & Constructors
Pittsburgh, Pa 15222

ELECTRIC MOTORS

Introduction

Today more mechanical equipment is driven by electric motors than all other drivers combined. According to the Federal Energy Administration electric motors consume over 75 percent of the electric power used by industry.

All motors are built to general acceptable industry standards. These motor standards are the National Electric Manufacture Association (NEMA), The Institute of Electrical and Electronics Engineers (IEEE), American National Standard Institute (ANSI) and the National Electrical Code (NEC). These standards set motor dimensions, tolerances, performances, temperature rise, types of insulation, test codes, etc.

Types of Motors

Motors are manufactured with the following enclosures. Open, drip-proof rent motors are divided into two types — induction or synchronous. Most motors in petroleum and chemical plants are induction type which drive pumps, fans, blowers, compressors, conveyors, machine tools, etc. and operate between 900 to 3,600 RPM. Synchronous motors have speeds of 600 RPM and below and relatively high horsepowers, 500 HP and up, such as drives for reciprocating compressors, or slow speed moving equipment. For low speeds cost favors synchronous motors over induction motors.

Squirrel cage induction motors 250 HP and below have been standardized by the National Electric Manufacture Association (NEMA). Prior to 1972 these horizontal motors were called NEMA "U" frame motors. Today horizontal motors are built to a more compact and economical design and are called "T" frame motors.

The "T" frame motors have standard motor dimensions so that different manufacturer's motors may be substituted for others. Some users still prefer the old "U" frame motors which are manufactured up to approximately 150 HP.

Direct current motors are specially designed where variable speeds (0 to 100 percent) are required. By changing the voltage or the strength of the magnetic field the speed of the D.C. motor is readily controlled.

171

Motor Enclosures

Motors are manufactured with the following enclosures. Open, drip-proof (ODP) motors are open type motors and used indoors in a non-hazard, non-corrosive atmosphere. Totally enclosed, fan cooled (TEFC) or totally enclosed, non-ventilated (TENV) motors are enclosed and completely protected. See Figure 5.1. TENV motors are manufactured normally up to 1½ HP and TEFC from 2 to 250 HP. However, some manufacturers do build TEFC motors up to 1,000 HP. TEFC motors have a fan in the end bell to circulate cooling air across the motor surfaces. These motors are the "workhorse" for industry including steel mills, refineries, chemical plants, etc. and are manufactured in two distinct lines. Standard TEFC line is suitable for outdoors in a relatively clean atmosphere. The severe-duty or mill and chemical (MAC) type, corro-duty type is another manufacturer's name, is designed for outdoor in a corrosive, dusty condition, such as chemical plant services.

General Purpose TEFC Motor   CORRO-DUTY Motor

Figure 5.1. A family of totally enclosed motors (60 Hertz).

Weather Protected I (WP-1) provides minimum outdoor protection from the rain, snow, and light dust. Ventilated passages can be provided with screen for protection from large objects, rodents, snakes, etc. These motors are available in sizes above 250 HP.

Weather Protected II (WP-II) are suitable for outdoor service where heavy rains, high winds, snow, and dust storms are frequent and available above 250 HP. The ventilating passages are arranged so that the rain and air-borne particles blown into the motor make at least three abrupt changes in directions of 90° and are discharged without entering the internal ventilating passages which lead to the electrical parts of the machine. See Figure 5.2 for picture of this motor.

Explosion-proof motors are specially designed TEFC or TENV motors to

Figure 5.2. Horizontal motors Weather Protected Type II.

withstand an internal explosion without igniting a flammable mixture outside of the motor and are suitable for hazardous locations (CLass I, Division I). They are manufactured in NEMA frames up to 250 HP. For motors above 250 HP a force-ventilated motor can be provided. Clean air is brought in from a safe location by a fan and blown across the motor for cooling and discharged to the outside. Since the system is pressurized, no contaminated gas can enter the motor. Necessary control must be provided on the system to start and shutdown the motor.

Totally Enclosed Water-Air Cooled (TEWAC) motors or Totally Enclosed Air-Air (TEAAC) motors are designed to operate under adverse environments, such as extreme corrosive chemical plants, where cooling with outside air is undesirable. These motors are cooled through an exchanger. As in the case of a TEAAC motor they are externally cooled through a built-in air to air heat-exchanger, see Figure 5.3. External air is drawn in and forced through the heat exchanger by means of a fan which is mounted on the motor shaft. The internal air is recirculated through the heat exchanger and then through the motor to pick up the heat generated by the electrical parts. These motors are available in sizes 250 HP and larger.

Voltages

Depending on the user's plant standard, voltage requirements may vary. Many chemical plants utilize 120 volts and 1 phase for motors ½ HP and below, 460 volts and 3 phase for ¾ HP to 200 HP motors, and 2,300 volts and 3 phase for 200 HP to 2,000 HP. For larger size motors, 4,160, 6,600 or 13,200 volts are generally available.

EXTERNAL AIR FLOW THRU BUNDLES

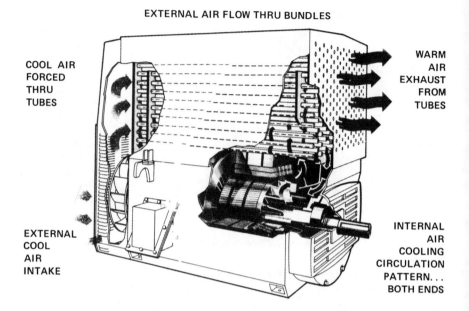

COOL AIR
FORCED
THRU
TUBES

WARM
AIR
EXHAUST
FROM
TUBES

EXTERNAL
COOL
AIR
INTAKE

INTERNAL
AIR
COOLING
CIRCULATION
PATTERN...
BOTH ENDS

Figure 5.3. Horizontal motors totally enclosed tube cooled.

Figure 5.4 shows relative cost of motors versus different voltages and also indicates the standard voltages as manufactured by one large motor producer.

Motor Selection & Pricing

Most customers have their own motor specifications. However, in selecting the type of motors, available power supply, horsepower rating, type of driven equipment and the ambient conditions must be considered. If a user does not have a motor standard, most motor manufacturers have data sheets which may be filled out for them to recommend a motor for the service.

Figure 5.5 plots dollar/HP versus motor HP for various types of squirrel-cage, induction motors, operating at 1,800 RPM and provides relative cost for different types of enclosures, as discussed above.

Recently motor manufacturers have designed a high-efficiency driver to increase its efficiency by 2 to 5 percent. The additional cost for these motors can be recovered in less than a year, (8,000 hours) due to the electric power savings, assuming a power cost of 3 to 4 cents/KWH. These motors are now available in NEMA frame enclosures from 1 to 250 HP. Some manufacturers call these motors high efficiency, mill and chemical type.

In selecting a motor the starting torque and current must be matched with the driven equipment. Motors are divided into the following most common NEMA classifications:

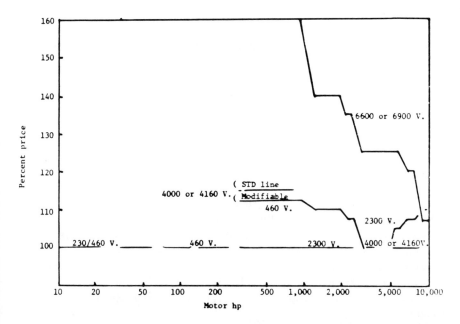

Figure 5.4. Cost vs voltage for motors from 10 to 10,000 hp. For modifiable designs, 250 hp and below other standard voltages such as 575 V. are available at the base price.

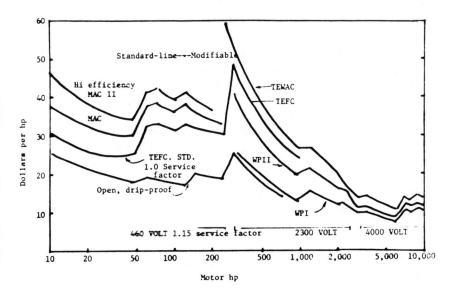

Figure 5.5. Motor prices in dollars per horsepower for 1,800 rpm, squirrel-cage, induction motors from 10 to 10,000 hp. Except for TEFC, Std. motors shown from 10 to 250 hp have service factor; above 250 hp the standard service factor is 1.0. Basis of this data is June, 1979.

NEMA A — High starting current and normal starting torque — used for high peak running loads.

NEMA B — Low starting current and normal torque — used for general purpose such as centrifugal compressors and pumps, blowers and fans, etc.

NEMA C — Low starting current and high starting torque — designed for reciprocating compressors and pumps, heavy loads, etc.

NEMA D — Lowest starting current and high starting torque — for flywheel and reversing duty.

Types of Turbines

The steam turbine is the next most widely used driver for process equipment. As a driver, the steam turbine has had more of an application in the petroleum industry than the chemical industry because the oil refiners normally require greater horsepower and higher speeds. Also, the exhaust steam is usable in the refining process and the steam turbine is considered more reliable than the local electrical power supply. However, today's chemical plants are being built considerably larger and there is now a trend to steam turbine driven equipment.

Horsepowers of single stage turbines range from small sizes, 5 to 10 HP, to a maximum of 3,000 HP with speed from 1,000 to 7,000 RPM. 3,550 RPM is the most common speed because in many cases the turbines drive centrifugal pumps which have an electric motor driven spare at the same speed. Steam turbines operate more efficiently at higher speeds. Their efficiency is directly proportional to speed. Depending on the turbine size, at 3,550 RPM the steam rate is approximately ½ as a turbine operating at 1,750 RPM. Operating turbine driven equipment below 2,000 RPM normally will justify the cost of a speed reducer with a higher speed turbine. A gear reducer will absorb 2 to 3 percent of the shaft horsepower.

The rotating movement of the turbine is produced by steam impinging on the buckets or blades which are mounted on a rotating wheel. The turbine wheels for single stage turbines are commonly manufactured with diameters from 9" to 28". Single stage turbines are usually designed to operate as noncondensing units where lower pressure steam can be used in the process or for building heat and are prime movers for pumps, fans, compressors, line shafts, paper machines, and small generators. See Figure 5.6 for cross-section of a single stage turbine.

High back pressure turbines are designed for high initial steam pressure and exhaust pressures, such as 650 psig inlet and 150 psig exhaust, and may have one or more stages. Common sizes are 150 to 3,000 HP with speeds from 5,000 to 10,000 RPM. These turbines generally drive pumps, fans, compressors, etc... and where there is a need for high pressure exhaust steam for the process.

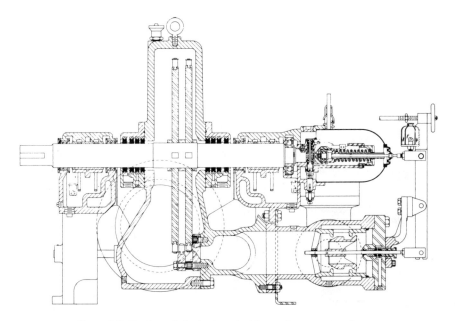

Figure 5.6. Section of single-stage turbine and governor system.

Medium multistage turbines have their applications in installations requiring higher efficiencies, and therefore have lower steam rates than are obtainable with single stage turbines. Operating cost or possibly existing boiler plant capacity may establish the required turbine efficiency. These turbines are available in sizes from 750 to 5,000 HP and speeds up to 10,000 RPM.

Large multistage condensing turbines are normally used for drives on large centrifugal or axial flow compressors, very large horizontal pumps, large electrical generators, etc. Horsepower of these turbines are in the 5,000 to 60,000 HP range and with speeds from 3,000 to 16,000 RPM. See Figure 5.7 for cross-section of a multistage turbine.

Steam Rates

Steam rate (or water rate) is a term which is exprssed as # steam/HP-HR or # steam/KW-HR and may be either theoretical or actual.

The theoretical steam rates are available from the turbine manufacturers or may be calculated from the Mollier Chart. See Figure 5.8.

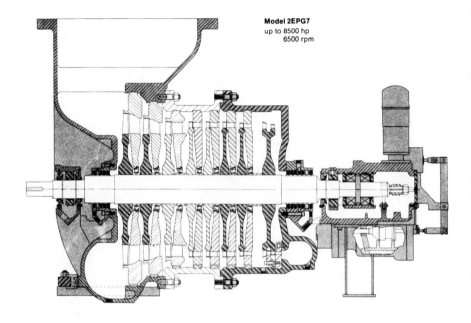

Model 2EPG7
up to 8500 hp
6500 rpm

Figure 5.7. Section of multistage turbine.

Example 1

Problem: Assume a single stage steam turbine with an inlet steam condition of 175 psig, 50 °F superheat; exhaust 20 psig. Calculate the theoretical available heat drop and theoretical steam rate.

Solution: See Figure 5.8 Mollier Diagram. Assume h_1 as enthalpy at inlet and h_2 as enthalpy at exhaust.

Theoretical available heat drop $= h_1 - h_2 = 1229 - 1094 = 135$ Btu/# steam

$$\text{Theoretical steam rate} = \text{TSR} = \frac{2545 \quad \dfrac{(BTU)}{(HP \cdot HR)}}{h_1 - h_2 \quad \dfrac{(BTU)}{(\# \text{ steam})}}$$

2545 BTU/HR is the heat to produce 1 HP

$$\text{TSR} = \frac{2545}{135} = 18.85 \ \frac{\# \text{ steam}}{(HP \cdot HR)} \quad \text{or}$$

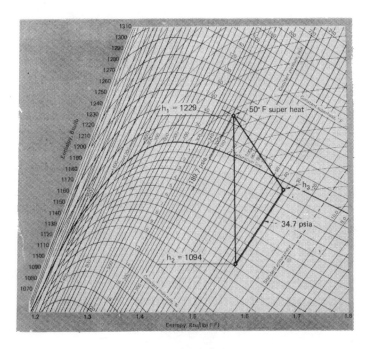

Figure 5.8. Mollier diagram gives theoretical available heat drop.

$$TSR = \frac{18.85}{.75\,\dfrac{KW}{HP}} = 25.27\,\frac{\#\ steam}{KW \cdot HR}$$

$$\text{The actual steam rate} = SR = \frac{\text{Theoretical Steam Rate}}{\text{Efficiency}} = \frac{TSR}{E}$$

The magnitude of the actual steam rate is determined by steam inlet conditions, exhaust pressure, turbine wheel diameter, number of stages, speed (RPM), pressure drop through inlet and outlet turbine parts, windage loss, gland leakage, and the mechanical efficiency of the turbine.

The actual steam rate is guaranteed by the manufacturer. Actual steam rate information is readily available in the catalogs of the different turbine manufacturers. Each manufacturer has his own method to calculate the actual steam rates. The curves in Figures 5.9 (a) to 5.9 (e), 5.10 (a) to 5.10 (e), and 5.11 can be used to calculate steam rates for single stage turbines with 14″, 18″, 22″ and 28″ wheel sizes in the following formula and as shown in Example #2 with an accuracy of ± 5%.

$$\text{Actual Steam Rate} = SR = \frac{\text{Base Steam Rate}}{\text{Superheat Factor}} \times \frac{HP + HP\ Loss}{HP}$$

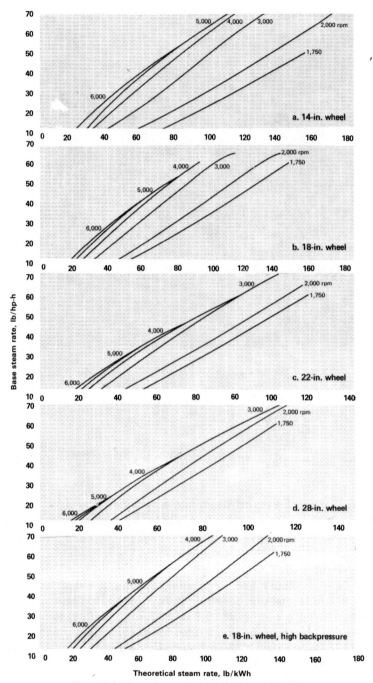

Figure 5.9. Base steam rates for single-stage turbines.

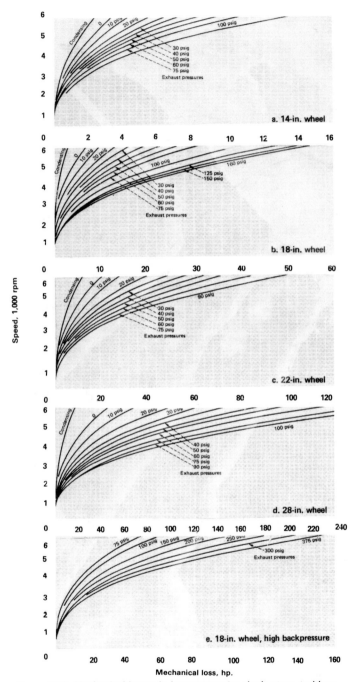

Figure 5.10. Mechanical losses in horsepower for single-stage turbines.

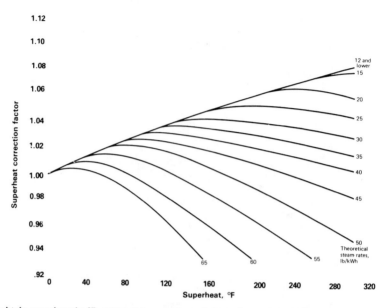

To obtain superheat in °F, subtract temperature of dry and saturated steam given in tabulation from total initial temperature.

Psig	Saturation, temperature, °F	Psig	Saturation, temperature, °F	Psig	Saturation, temperature °F	Psig	Saturation, temperature, °F
0	212	150	366	300	422	450	460
5	228	155	368	305	423	455	461
10	240	160	371	310	425	460	462
15	250	165	373	315	426	465	463
20	259	170	375	320	428	470	464
25	267	175	378	325	429	475	465
30	274	180	380	330	431	480	466
35	281	185	382	335	432	485	467
40	287	190	384	340	433	490	468
45	293	195	386	345	434	495	469
50	298	200	388	350	436	500	470
55	303	205	390	355	437	510	472
60	312	215	394	365	440	530	476
65	423	215	394	365	440	530	476
70	316	220	396	370	441	540	478
75	320	225	408	375	442	550	480
80	328	230	399	380	444	560	482
85	328	235	401	385	445	570	483
90	331	240	403	390	446	580	485
95	335	245	404	395	447	590	487
100	338	250	406	400	448	600	489
105	341	255	408	405	449	610	491
110	344	260	410	410	451	620	492
115	347	265	411	415	452	630	494
120	350	270	413	420	453	640	496
125	353	275	414	425	454	650	497
130	356	280	416	430	455	660	499
135	358	285	417	435	456	670	501
140	361	290	419	440	457	680	502
145	364	295	420	445	458	690	504

Figure 5.11. Superheat correction factor from dry and saturated steam temperature.

Example 2

Problem: Assume the same steam conditions of Example #1, the single stage turbine is 500 HP with a 28" wheel and operating at 3,550 RPM. What is the actual steam rate, turbine efficiency, actual heat drop and actual exhaust temperature.

Solution: Base steam rate of a single stage turbine is determined by using Figure 5.9 (d), 28" Wheel. By knowing the theoretical steam rate in #/ KW · HR and speed,

$$TSR = 25.27 \frac{\text{\# steam}}{KW \cdot HR} \; ; \; Speed = 3,550 \; RPM$$

From Figure 5.9 (b) the base steam rate is 37 #/hr. HP loss is determined by using Figure 5.10 (d), 28 wheel. By knowing the speed of 3,550 RPM and exhaust pressure of 20 psig, HP loss from Figure 5.10 (d) is 22. Superheat factor is determined from Figure 5.11, knowing TSR of 25.27 and 50°F superheat. Superheat factor is 1.018.

.

$$SR = \frac{37}{1.018} \times \frac{500 + 22}{500} = 37.94 \; \frac{\#}{HP \cdot HR}$$

$$\text{Turbine Efficiency (E)} = \frac{\text{Theoretical Steam Rate (TSR)\#}}{\text{Actual Steam Rate (SR)}}$$

$$= \frac{18.85}{37.94} = .497 \text{ or } 49.7\%$$

Actual Heat Drop of the Steam $= (h_1 - h_2) \times E$

$$= 135 \times .497 = 67.10 \; \frac{BTU}{\text{\# steam}}$$

Actual Enthalpy of Exhaust Steam $= h_3$,

$h_3 = h_1 -$ Actual Heat Drop $= 1229 - 67.10 = 1161.90$ BTU/#STM

In looking at Figure 5.8 and locating h_3, h_3 falls near the 1 percent moisture for 34.7 psia or 20 psig. Steam temperature for 34.7 psia saturated steam is 259 °F.

Figures 5.9 (e) and 5.10 (e) may be used to calculate steam rates for single stage high back pressure turbines with 18″ wheel. This type of turbine is normally manufactured in limited wheel size of 16″, 18″, and 20″.

The efficiencies in Figures 5.12 (a) to 5.12 (d) and 5.13 are for medium and large size multistage turbines respectively, and are only approximate, however, they may serve the purpose in comparing efficiencies of the different types of turbines. Actual steam rates for these turbines may be calculated as described above.

Turbine Selection

Selecting the type of steam turbines for a plant or process unit should be made so that the steam requirements or the steam rates of the various types of operating turbines will provide a steam balance within the battery unit or plant.

A cost should be set for the quantity of steam used. Initial steam turbine or equipment cost must be considered. A single stage turbine with a 28″ wheel will have a higher initial cost than a turbine with a 18″ wheel (see Figure 5.14), but the steam rates are lower. A multistage turbine with the greatest number of stages will normally have a higher initial cost than a turbine with fewer stages, again the steam rate will be lower. Unless some other factors exist, such as steam balance, the selection between turbines of various initial costs and steam rates depends upon the payout period. Normally the turbine with the lower steam rate will justify the additional cost. Evaluation of the savings is illustrated in Example #3 below.

Example 3

Problem: Given the same 500 HP as Example #2 with 28″ wheel. Would it be more economical to use an 18″ wheel?

Solution: See Figure 5.14, Horsepower vs. Cost of Single Stage Turbines. Assume steam rate in Example #2 of 37.94, or say 38. The steam rate for an 18″ wheel calculates to be approximately 48.

Wheel Size	Cost of Turbine	SR		Steam Savings Steam Cost	
18″	$11,000	48	48 $\dfrac{\# \text{STM}}{\text{BHP} \cdot \text{HR}}$ × 500 HP × $6.0/1000#STM	= 144
28″	$12,700	38	38 $\dfrac{\# \text{Steam}}{\text{BHP} \cdot \text{HR}}$ × 500 HP × $6.0/1000#STM	= $\dfrac{114}{\$30/\text{hr}}$
DIFF.	$ 1,700				

$$\text{Payout Period} = \frac{\$1,700}{\$30/\text{hr}} = 56 \text{ Hours or approximately 2–3 Days}$$

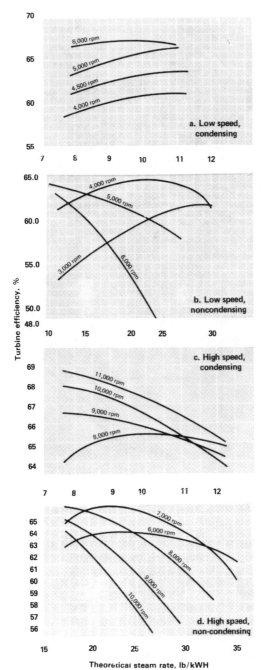

Figure 5.12. Approximate efficiencies for medium-size turbines (1,000 to 5,000 hp).

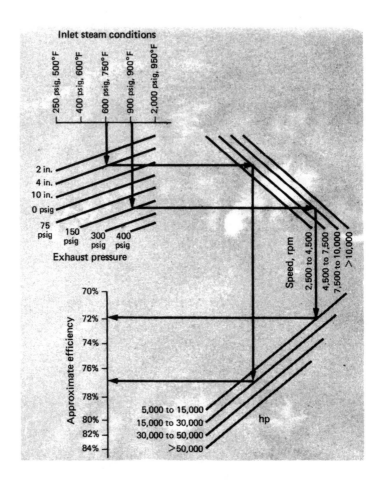

Figure 5.13. Approximate efficiencies for large-size turbines.

Figure 5.15 was plotted from information which was taken from several manufacturer catalogs and averaging these steam rates for the turbine conditions in Example #2. The steam rate calculated in Example #2 (37.94 #/HP · HR) agrees fairly accurately with Figure 5.15, 39 #/HP ·HR. In larger wheel sizes the steam rate is less which indicates better efficiency. However, each wheel size has horsepower limitations because of the quantity of steam which can be passed through the turbine. Figure 5.15 shows the horsepower limitation for each wheel size. For lower horsepowers these curves converge. At 100 HP the steam rate curve for the 28″ diameter wheel intersects the 25″ diameter wheel.

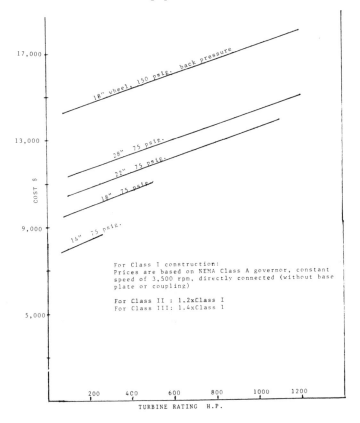

Figure 5.14. Approximate costs for single stage steam turbines (1980).

This windage loss for the larger wheel has increased to such an extent as to become less efficient than the smaller diameter.

Evaluation of multistage turbines will require the manufacturer's assistance, because pricing and steam rates vary considerably. This is especially true for condensing multistage turbines, since the condenser and steam piping cost may require consideration. Most turbine manufacturers are always ready and willing to simplify the task of choosing the best turbine for the application, since these options have been evaluated by their computers.

Steam rate or the quantity (#/HR) of steam available should be specified since it is as important as other steam turbine requirements because cost and size of turbines are affected.

OTHER DRIVES

Gas Turbines drive large electric generators and centrifugal compressors. One large applciation of the gas turbine is for peaking electrical loads especially

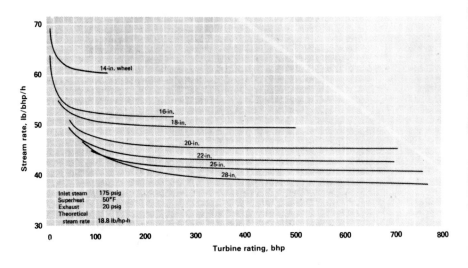

Figure 5.15. Steam rates for single-stage, 3,550-rpm turbines with 14- to 28-inc. wheel sizes.

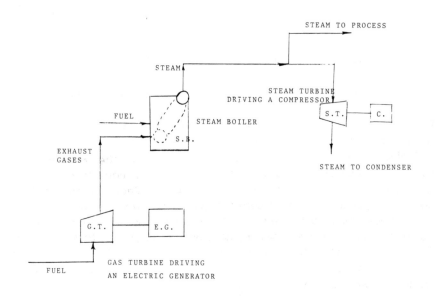

Figure 5.16. Simple gas turbine cycle (recovering heat from exhaust gases in a waste heat steam boiler).

in central power stations where electrical loads are changing rapidly. The unit can be started, put on the line, operated at full load in a matter of a few minutes. Another use is in gas fields and pipe line stations where natural gas is plentiful and relatively cheap. Natural gas, process gases, light liquid fuels, LPG and even coal may be used as fuels. Gas turbines are manufactured from 1,000 to 30,000 HP. The overall efficiency of the gas turbine may be increased considerably from approximately 25 percent to over 70 percent by installing a waste heat boiler to utilize the heat from the turbine exhaust. The steam from the boiler may be used in the process, to drive turbines, or heating buildings. See Figure 5.16 as a typical waste heat boiler cycle.

Diesel, gas or steam engines generally drive reciprocating compressors, emergency or stand-by generators, pumps, etc. Both diesel and gas engine driven generators serve as stand-by power for industries, commercial complexes, institutions, small power stations. Generally, the horsepowers range from 50 to 5,000 HP. Again the heat in the exhaust gases from gas or diesel engine may be recovered by exhausting gases through a heat exchanger to increase the overall engine efficiency from 40 percent to over 70 percent. The following table indicates the types of reciprocating compressors which are generally driven by these engines.

Table 5.1. Drives Used for Reciprocating Compressors

Type of Reciprocating Compressors	Driver Generally Used	Approximate HP — Ranges
Horizontal — Straight Line	Steam Engine	15 — 150
Horizontal — Four Center	Steam Engine	150 — 2,000
Angle Types — Horizontal	Vertical Gas,	200 — 3,000
Compressor Cylinder	Steam, Diesel	
	Cylinders	

ACKNOWLEDGEMENT AND REFERENCES

1. Steam Rate Curves taken from Elliott Turbine Bulletins — H-3F, "Single-Stage Turbines," and H-36, "Multi-Stage Turbines," Figure 5-8 through Figure 5.14.
2. Figure 5.4, Percent Price vs. Motor HP, and Figure 5.5, Dollar/HP vs. Motor HP, taken from C. S. Olson's article "Select Motors To Save Energy" in Hydrocarbon Processing, August 1979.
3. Turbine information taken from W. W. Willoughby's article, "Steam Rate" in Chemical Engineering, September 11, 1979.
4. Pictures of Motors in Figures 5.1 to 5.3 provided by U. S. Motors.
5. Sections of Turbine in Figure 5.6 and Figure 5.7 provided by Elliott Turbine.

CHAPTER 6

PIPING & VALVES — DESIGN, CODES & APPLICATIONS

J. L. HARRIS
Hooker Chemical Co.
Grand Island, N.Y.

INTRODUCTION

The intent of this chapter is to present an overview of the mechanical aspects of piping systems. The emphasis will be on aboveground piping systems in the moderate temperature-pressure range.

Undoubtedly some references may have been overlooked but not intentionally. References have been included in various sections that are considered typical. General references are included in the last section.

One source of information about pipes, valves and fittings which should not be overlooked is the manufacturers of these items. Manufacturers' literature has been used extensively in the preparation of this chapter.

The material presents the author's view of current piping practice. Should this material be at variance with ANSI B31.3 or any other piping code in situations where the codes govern, then the codes are to take precedence over the material presented here. The author makes no warranty of any nature concerning such material, industrial standards or practices.

STANDARDS AND SPECIFICATIONS

Published standards and specifications have been cited extensively in this chapter. The following organizations have developed standards and specifications relating to piping systems, materials and fabrication.

American National Standards Institute (ANSI)
American Petroleum Institute (API)
American Society of Mechanical Engineers (ASME)
American Society for Testing & Materials (ASTM)
American Welding Society (AWS)
Manufacturers Standardization Society of the Valve and Fitting Industry (MSS)
National Bureau of Standards (NBS)
Pipe Fabrication Institute (PFI)

PIPE SIZING

Sizing the piping system starts with establishing the rate of flow. The next step is to choose a trial line size. One method is to assume a size. A second method is to use a suggested velocity, as shown in Table 6.1 to calculate a line size or range of sizes (1).

Table 6.1. Suggested Fluid Velocities

The velocities are suggestive only and are to be used to approximate line size as a starting point for pressure drop calculations. The final line size should be such as to give an economical balance between pressure drop and reasonable velocity.

Fluid	Suggested Trial Velocity	Pipe Material
Acetylene (Observe pressure limitations)·	4000 fpm	Steel
Air, 0 to 30 psig	4000 fpm	Steel
Ammonia		
Liquid	6 fps	Steel
Gas	6000 fpm	Steel
Benzene	6 fps	Steel
Bromine		
Liquid	4 fps	Glass
Gas	2000 fpm	Glass
Calcium Chloride	4 fps	Steel
Carbon Tetrachloride	6 fps	Steel
Chlorine (Dry)		
Liquid	5 fps	Steel, Sch. 80
Gas	2000-5000 fpm	Steel, Sch. 80
Chloroform		
Liquid	6 fps	Copper & Steel
Gas	2000 fpm	Copper & Steel
Ethylene Gas	6000 fpm	Steel
Ethylene Dibromide	4 fps	Glass
Ethylene Dichloride	6 fps	Steel
Ethylene Glycol	6 fps	Steel
Hydrogen	4000 fpm	Steel
Sodium Hydroxide		
0-30 Percent	6 fps	Steel
30-50 Percent	6 fps	Steel and Nickel
50-73 Percent	4	Nickel
Sodium Chloride Sol'n.		
No Solids	5 fps	Steel
With Solids	(6 Min.-15 Max.)	Monel or nickel
Perchloroethylene	7.5 fps	
	6 fps	Steel
Steam		
0-30 psi Saturated*	4000-6000 fpm	Steel
30-150 psi Saturated or superheated*	6000-10000 fpm	
150 psi up superheated	6500-15000 fpm	
*Short lines	15,000 fpm (max.)	
Sulfuric Acid		
88-93 Percent	4 fps	S.S.-316, Lead
93-100 Percent	4 fps	Cast Iron & Steel Sch. 80

(Continued)

Table 6-1. Suggested Fluid Velocities (Continued)

Fluid	Suggested Trial Velocity	Pipe Material	Fluid	Suggested Trial Velocity	Pipe Material
Hydrochloric Acid			Sulfur Dioxide	4000 fpm	Steel
Liquid	5 fps	Rubber Lined	Styrene	6 fps	Steel
Gas	4000 fpm	R.L., Saran, Haveg	Trichloroethylene	6 fps	Steel
			Vinyl Chloride	6 fps	Steel
Methyl Chloride			Vinylidene Chloride	6 fps	Steel
Liquid	6 fps	Steel	Water		
Gas	4000 fpm	Steel	Average service	3-8 (avg. 6) fps	Steel
Natural Gas	6000 fpm	Steel	Pump suction lines	3-8 fps	Steel
Oils, lubricating	6 fps	Steel	Maximum economical (usual)	7-10 fps	Steel
Oxygen			Sea and brackish water, lined pipe	5-8 fps 3	R.L., concrete, asphalt-line, saran-lined, transite
(ambient temp.)	1800 fpm Max.	Steel (300 psig Max.)	Concrete	5-12 fps (Min.)	
(Low temp.)	4000 fpm	Type 304 SS			
Propylene Glycol	5 fps	Steel			

Note: R.L. = Rubber-lined steel.

(Concluded)

With the flow rate and line size determined, the pressure drop can be read from a table, determined from a set of nomographs or calculated. For convenience, the answer is usually in terms of pounds per square gage or feet of flowing fluid per 100 feet of pipe (psi/100' or ft/100') (Table 6.2 and 6.3).

To determine the total length of piping in a system the number of fittings and valves must be taken into consideration. Fittings and valves can be expressed in terms of equivalent feet of pipe and added to the feet of pipe for a total equivalent length of pipe on which to base the pressure loss calculation (Table 6.4).

An alternate method for finding the total pressure loss in the system is to calculate the pressure loss in the pipe, valves and fittings separately and add the values. The loss through the fittings and valves can be calculated using the appropriate k values (resistance coefficients).

To compensate for what cannot be taken into account when calculating the pressure loss the addition of 10 to 15% to the calculated figure is suggested to arrive at a total pressure loss figure.

PIPE SYSTEMS

Suggested piping systems for different materials is illustrated in Tables 6.5 through 6.8. Although this is not an exhaustive compilation, it represents what can be considered as acceptable usages.

PIPE

Standard Sizes

Metallic and most extruded plastic pipe for aboveground pressure service are manufactured to the same set of dimensional standards. These dimensions are shown in Table 6.10, and are a combination of the data in ANSI B36.10 and B36.19. It is important to note that not all sizes are available in all of the materials of which pipe can be manufactured. Table 6.11 details the pipe size limitations and applicable specifications.

The outside dimensions of plastic lined steel pipe are the same as those of unlined steel. The difference is in a smaller inside diameter due to plastic lining.

Pressure Rating

The pressure capability of pipe depends on pipe diameter, wall thickness and the allowable stress or hydrostatic design stress of the material involved. These latter values can be found in ANSI B31.3, Appendix A.

Generally, the pressure capability of metallic pipe is more than sufficient for moderate temperatures and pressures. In the event the pressure capability of the pipe is to be determined the method that should be used is shown in ANSI

Table 6.2. Pressure Drop of Water Through Schedule 40 Steel Pipe

G.P.M.	FT³PerSec.	1/8" V Ft/Sec	1/8" P psi	1/4" V	1/4" P	3/8" V	3/8" P	1/2" V	1/2" P	3/4" V	3/4" P	1" V	1" P	1 1/4" V	1 1/4" P	1 1/2" V	1 1/2" P	2" V	2" P	2 1/2" V	2 1/2" P	3" V	3" P	3 1/2" V	3 1/2" P	4" V	4" P	5" V	5" P	6" V	6" P	8" V	8" P	10" V	10" P
.1	.00022	.56	.677																																
.2	.00045	1.14	2.48	.62	.548																														
.3	.00067	1.70	5.26	.93	1.16	.50	.255																												
.4	.00089	2.26	9.00	1.24	1.98	.67	.436	.42	.136																										
.5	.00111	2.82	13.58	1.55	3.00	.84	.656	.53	.205	.30	.050																								
.6	.00134	3.38	19.12	1.85	4.22	1.01	.925	.63	.290	.36	.071																								
.8	.00178	4.52	32.62	2.47	7.17	1.34	1.58	.84	.494	.48	.121	.30	.036																						
1	.00223			3.09	10.91	1.68	2.39	1.06	.749	.60	.183	.37	.055	.21	.014																				
2	.00446			6.18	39.60	3.36	8.68	2.11	2.72	1.20	.665	.74	.199	.43	.051																				
3	.00668					5.04	18.46	3.17	5.77	1.80	1.41	1.11	.424	.64	.107																				
4	.00891					6.72	31.55	4.22	9.86	2.40	2.42	1.49	.724	.86	.183																				
5	.01114							5.28	14.92	3.01	3.64	1.86	1.09	1.07	.276																				
6	.01337							6.33	20.95	3.61	5.13	2.23	1.54	1.29	.390																				
8	.01782									4.81	8.76	2.97	2.62	1.71	.667	1.26	.308																		
10	.02228									6.01	13.28	3.713	3.97	2.142	1.01	1.58	.466																		
15	.03342											5.57	8.46	3.21	2.14	2.36	.992	1.43	.285																
20	.04456											7.43	14.42	4.28	3.66	3.15	1.69	1.91	.486																
25	.05570													5.36	5.54	3.94	2.54	2.39	.736																
30	.06684													6.43	7.79	4.73	3.60	2.87	1.03	2.01	.424														
35	.07798													7.50	10.38	5.51	4.79	3.35	1.37	2.35	.566														
40	.08912													8.57	13.28	6.30	6.14	3.82	1.76	2.68	.724														
50	.1114															7.88	9.31	4.78	2.67	3.35	1.10	2.17	.371												
60	.1337															9.45	13.08	5.74	3.75	4.02	1.54	2.61	.520												
70	.1560																	6.70	4.99	4.70	2.05	3.04	.693	2.27	.335										
80	.1782																	7.65	6.40	5.37	2.63	3.47	.890	2.59	.430										
90	.2005																	8.60	7.96	6.04	3.28	3.91	1.10	2.92	.535										
100	.2228																	9.56	9.69	6.71	3.98	4.34	1.34	3.24	.650	2.52	.346								
125	.2785																			8.38	6.03	5.43	2.01	4.05	.984	3.15	.523								
150	.3342																			10.1	8.46	6.52	2.86	4.87	1.38	3.78	.734								
175	.3899																			11.7	11.3	7.60	3.81	5.68	1.84	4.41	.978	2.81	.316						
200	.4456																			13.4	14.4	8.69	4.89	6.49	2.36	5.04	1.25	3.21	.405						
225	.5013																					9.77	6.09	7.30	2.94	5.67	1.56	3.61	.505						
250	.5570																					10.9	7.41	8.11	3.58	6.30	1.90	4.01	.616	2.78	.245				
275	.6127																					11.9	8.84	8.92	4.27	6.93	2.27	4.41	.734	3.06	.292				
300	.6684																					13.0	10.4	9.73	5.02	7.56	2.67	4.81	.863	3.33	.344				
350	.7798																					15.2	13.8	11.4	6.87	8.82	3.55	5.62	1.15	3.89	.457				
400	.8912																							13.0	8.58	10.1	4.56	6.41	1.47	4.44	.587	2.57	.149		
450	1.003																											7.22	1.83	5.00	.731	2.89	.185		
500	1.114																											8.02	2.23	5.55	.887	3.21	.225		
550	1.225																											8.82	2.67	6.11	1.07	3.53	.270		
600	1.337																											9.62	3.13	6.66	1.25	3.85	.316		

Table 6.2. Pressure Drop of Water Through Schedule 40 Steel Pipe (Continued)

G.P.M.	FT Per Sec	4" V	4" P	5" V	5" P	6" V	6" P	8" V	8" P	10" V	10" P	12" V	12" P	14" V	14" P	16" V	16" P	18" V	18" P
700	1.560	17.6	12.9	11.2	4.16	7.78	1.66	4.49	.420	2.85	.135								
750	1.671	18.9	14.7	12.0	4.75	8.33	1.89	4.81	.480	3.05	.154								
800	1.782	20.2	16.5	12.8	5.35	8.89	2.13	5.13	.540	3.26	.173								
850	1.894	21.4	18.5	13.6	5.98	9.44	2.38	5.45	.605	3.46	.194								
900	2.005	22.7	20.6	14.4	6.65	10.0	2.66	5.77	.627	3.66	.216	2.58	.090						
950	2.117	23.9	22.8	15.2	7.36	10.6	2.93	6.09	.744	3.87	.238	2.72	.099						
1000	2.228			16.0	8.10	11.1	3.23	6.41	.817	4.07	.262	2.87	.109						
1100	2.451			17.6	9.66	12.2	3.85	7.06	.975	4.48	.313	3.15	.130						
1200	2.674			19.2	11.4	13.3	4.53	7.70	1.15	4.88	.368	3.44	.153	2.85	.096				
1300	2.896			20.8	13.2	14.4	5.26	8.34	1.33	5.29	.427	3.73	.178	3.08	.111				
1400	3.119			22.4	15.1	15.6	6.01	8.98	1.53	5.70	.490	4.01	.204	3.32	.127				
1500	3.342			24.1	17.2	16.7	6.84	9.62	1.74	6.10	.556	4.30	.232	3.56	.145				
1600	3.565					17.8	7.73	10.3	1.96	6.51	.628	4.59	.262	3.79	.163	2.91	.084		
1800	4.010					20.0	9.64	11.5	2.46	7.32	.782	5.16	.329	4.27	.203	3.27	.104		
2000	4.456					22.2	11.6	12.8	2.97	8.14	.953	5.73	.396	4.74	.247	3.63	.127		
2500	5.570					27.8	17.6	16.0	4.49	10.2	1.44	7.17	.601	5.93	.374	4.54	.192	4.30	.149
3000	6.684							19.2	6.30	12.2	2.02	8.60	.842	7.11	.525	5.45	.270	5.02	.199
3500	7.798							22.4	8.41	14.2	2.70	10.0	1.12	8.30	.700	6.36	.358	5.74	.255
4000	8.912							25.7	10.8	16.3	3.46	11.5	1.44	9.48	.896	7.26	.459	6.45	.317
4500	10.03							28.9	13.4	18.3	4.31	12.9	1.76	10.7	1.12	8.17	.671		
5000	11.14									20.4	5.23	14.3	2.18	11.9	1.36	9.08	.695	7.17	.386
6000	13.37									24.4	7.35	17.2	3.06	14.2	1.91	10.9	.977	8.60	.542
7000	15.60									28.5	9.80	20.1	4.08	16.6	2.54	12.7	1.30	10.0	.723
8000	17.82											22.9	5.22	19.0	3.25	14.5	1.67	11.5	.926
9000	20.05											25.8	6.51	21.3	4.06	16.3	2.08	12.9	1.15
10000	22.28											28.7	7.91	23.7	4.92	18.2	2.53	14.3	1.40
12000	26.74													28.5	6.92	21.8	3.55	17.2	1.97
14000	31.19															25.4	4.72	20.1	2.62
16000	35.65															29.1	6.06	22.9	3.36
18000	40.10															32.7	7.55	25.8	4.18
20000	44.56																	28.7	5.08

Based on Saph and Schoder Formula $\Delta P = \dfrac{LQ^{1.86}}{1435\, d^5}$

By comparing physical properties, one can approximately estimate pressure drop for other liquids as well.

(Concluded)

Table 6.3. Pressure Drop of Air Thorough Schedule 40 Steel Pipe

FLOW Cu. Ft. per Min.

Free Air 1 atm. 60°F.	Comp. Air 100 psi Gage 60°F.	1/8" & 1½" V Ft/Sec	1/8" & 1½" P Psi	1/4" & 2" V Ft/Sec	1/4" & 2" P Psi	3/8" & 2½" V Ft/Sec	3/8" & 2½" P Psi	1/2" & 3" V Ft/Sec	1/2" & 3" P Psi	3/4" & 3½" V Ft/Sec	3/4" & 3½" P Psi	1" & 4" V Ft/Sec	1" & 4" P Psi	1¼" V Ft/Sec	1¼" P Psi
1	.128	5.42	.418	2.96	.092	1.61	.020								
2	.256	10.84	1.52	5.94	.325	3.22	.074								
3	.384	16.26	3.22	8.91	.711	4.82	.157								
4	.513	21.68	5.51	11.85	1.21	6.43	.266								
5	.641	27.10	8.30	14.81	1.84	8.04	.403	5.07	.126						
6	.769			17.78	2.59	9.65	.568	6.08	.177						
8	1.025			23.70	4.40	12.86	.977	8.10	.302	4.61	.074				
10	1.282			29.63	6.67	16.07	1.47	10.13	.457	5.76	.112				
15	1.922					24.11	3.12	15.20	.980	8.64	.239	5.34	.072		
20	2.563					32.14	5.30	20.29	1.67	11.52	.407	7.12	.122		
25	3.204					40.18	8.03	25.32	2.52	14.40	.617	8.90	.185		
30	3.845							30.39	3.54	17.28	.866	10.68	.260	6.16	.066
35	4.486							35.46	4.71	20.16	1.16	12.46	.346	7.19	.088
40	5.126							40.52	6.04	23.04	1.48	14.24	.443	8.22	.112
50	6.408							50.65	9.15	28.80	2.24	17.81	.671	10.27	.170
60	7.690	9.07	.110							34.55	3.30	21.37	.976	12.32	.239
70	8.971	10.58	.147							40.31	4.19	24.93	1.25	14.38	.318
80	10.25	12.09	.189							46.07	5.36	28.49	1.61	16.43	.407
90	11.53	13.60	.234	8.25	.067					51.83	6.69	32.05	2.00	18.49	.507
100	12.82	15.11	.286	9.17	.082							35.61	2.44	20.54	.617
125	16.02	18.89	.432	11.46	.124							44.51	3.69	25.68	.935
150	19.22	22.67	.607	13.76	.174							53.42	5.17	30.81	1.31
175	22.43	26.44	.810	16.05	.232	11.26	.096					62.32	6.90	35.95	1.75
200	25.63	30.22	1.035	18.34	.298	12.86	.122					71.22	8.84	41.08	2.24
225	28.84	34.00	1.29	20.63	.370	14.47	.152					80.12	11.01	46.22	2.79
250	32.04	37.78	1.57	22.93	.451	16.08	.185							51.35	3.40
275	35.24	41.55	1.88	25.22	.539	17.69	.222							56.49	4.06
300	38.45	45.33	2.21	27.51	.632	19.30	.260	12.50	.088					61.62	4.76
350	44.86	52.89	2.94	32.10	.843	22.51	.346	14.58	.117					71.89	6.34
400	51.26	60.44	3.76	36.68	1.08	25.73	.445	16.66	.150					82.16	8.14
450	57.67	68.00	4.68	41.27	1.34	28.94	.553	18.74	.187	14.00	.090				
500	64.08	75.55	5.70	45.85	1.63	32.16	.673	20.83	.227	15.55	.110				
550	70.49	83.11	6.82	50.44	1.96	35.38	.805	22.91	.272	17.11	.131				
600	76.90	90.66	8.02	55.02	2.30	38.59	.945	24.99	.319	18.66	.154	14.50	.082		
650	83.30	98.62	9.30	59.61	2.66	41.81	1.09	27.07	.370	20.22	.179	15.71	.095		

(Continued)

Table 6.3. Pressure Drop of air Through Schedule 40 Steel Pipe (Continued)

FLOW Cu. Ft. per Min. Free Air 1 atm. 60°F.	Comp. Air 100 psi Gage 60°F.	2" V Ft/Sec	2" P Psi	2½" V Ft/Sec	2½" P Psi	5" V Ft/Sec	5" P Psi	6" V Ft/Sec	6" P Psi	8" V Ft/Sec	8" P Psi	10" V Ft/Sec	10" P Psi	12" V Ft/Sec	12" P Psi
700	89.71	64.19	3.06	45.02	1.26										
750	96.12	68.78	3.48	48.24	1.43										
800	102.5	73.36	3.92	51.46	1.61										
900	115.3	82.53	4.88	57.89	2.01	13.83	.056								
1000	128.2	91.70	5.94	64.32	2.44	15.37	.068								
1100	141.0	100.9	7.07	70.75	2.92	16.91	.081								
1200	153.8	110.0	8.37	77.18	3.42	18.44	.096								
1300	166.6	119.2	9.67	83.62	3.98	19.98	.111								
1400	179.4			90.05	4.56	21.52	.128								
1500	192.2			96.48	5.19	23.05	.145	15.98	.058						
1600	205.1			102.9	5.85	24.59	.164	17.04	.065						
1800	230.7			115.8	7.36	27.66	.206	19.17	.082						
2000	256.3			128.6	8.86	30.74	.248	21.30	.099						
2500	320.4					38.43	.376	26.63	.150						
3000	384.5					46.11	.527	31.95	.210						
4000	512.6					61.48	.900	42.60	.359	24.60	.091				
5000	640.8					76.85	1.37	53.25	.544	30.75	.138				
6000	769.0					92.22	1.92	63.90	.765	36.90	.195	23.42	.062		
7000	897.1					107.6	2.55	74.55	1.02	43.05	.259	27.32	.083		
8000	1025.					123.0	3.27	85.20	1.31	49.20	.332	31.23	.106		
9000	1153.					138.3	4.08	95.85	1.63	55.35	.414	35.13	.132		
10000	1282.					153.7	4.96	106.5	1.97	61.50	.500	39.03	.160	27.48	.067
12000	1538.					184.4	6.96	127.8	2.78	73.80	.702	46.84	.226	32.98	.094
14000	1794.					215.2	9.26	149.1	3.69	86.10	.936	54.64	.301	38.47	.125
16000	2051.					245.9	11.88	170.4	4.74	98.40	1.21	62.45	.385	43.97	.160
18000	2307.					276.7	14.80	191.7	5.89	110.7	1.50	70.26	.479	49.46	.200
20000	2563.							213.0	7.17	123.0	1.82	78.06	.583	54.96	.242
22000	2820.							234.3	8.58	135.3	2.18	85.87	.696	60.46	.290
24000	3076.							255.6	10.1	147.6	2.56	93.67	.817	65.95	.341
26000	3332.							276.9	11.7	159.9	2.98	101.5	.950	71.45	.396
28000	3588.							298.2	13.4	172.2	3.41	109.3	1.09	76.94	.454
30000	3845.							319.5	15.3	184.5	3.88	117.1	1.24	82.44	.516
35000	4486.									215.3	5.13	136.6	1.65	96.18	.687
40000	5126.									246.0	6.58	156.1	2.12	109.9	.881
50000	6408.									307.5	9.97	195.2	3.21	137.4	1.33

Based on Fritzche's Formula $\Delta p = \dfrac{LQ^{1.84}}{1480\,P_c d^{6}}$

By comparing physical properties, one can approximately estimate pressure drop for other gases.

(Concluded)

Table 6.4. Equivalent Resistance of Valves and Fittings

K—FACTOR *		Gate Valve	Globe Valve	Angle Valve	45° Elbow	90° Elbow	180° Close Ret.	Tee Thru Run	Tee Thru Branch
		.22	10	5	.42	.90	2	.5	1.80
Nominal Pipe Size Inches	Inside Diameter Inches		L = Equivalent Length of Schedule 40 Pipe in Feet						
½	0.622	.41	18.5	9.3	.78	1.67	3.71	.93	3.33
¾	0.824	.54	24.5	12.3	1.03	2.21	4.90	1.23	4.41
1	1.049	.69	31.2	15.6	1.31	2.81	6.25	1.56	5.62
1¼	1.380	.90	41.0	20.5	1.73	3.70	8.22	2.06	7.40
1½	1.610	1.05	48.0	24.0	2.15	4.31	9.59	2.40	8.63
2	2.067	1.35	61.5	30.8	2.59	5.55	12.3	3.08	11.6
2½	2.469	1.62	73.5	36.8	3.09	6.61	14.7	3.68	13.2
3	3.068	2.01	91.5	45.8	3.84	8.23	18.2	4.57	16.4
4	4.026	2.64	120	60.0	5.03	10.8	23.9	6.00	21.6
5	5.047	3.30	150	75.0	6.31	13.5	30.0	7.51	27.0
6	6.065	3.98	180	90.0	7.10	16.2	36.1	9.05	32.5
8	7.981	5.23	237	118.5	10.0	22.4	47.5	11.9	42.8
10	10.02	6.56	298	149	12.5	26.8	59.6	14.9	53.7
12	11.94	7.83	356	178	15.0	32.0	71.1	17.8	64.0
14	13.13	8.60	392	196	16.4	35.2	78.1	19.5	70.4
16	15.00	9.84	446	223	18.8	40.2	89.3	22.4	80.5
18	16.88	11.1	502	251	21.2	45.2	100	25.2	90.5
20	18.81	12.3	560	280	23.5	50.5	112	28.0	101
24	22.63	14.8	672	336	28.3	60.6	135	33.7	121

(*Pressure drop can be expressed as $h = \dfrac{KV^2}{2g}$)

Table 6.5. Suggested Systems for Metallic Piping

System	Pipe	Fittings Screwed	Socket	Flanged	Butt Weld	Flange
.1	Steel (M) A-53 (M) A-106 (D) B36.10	Cast Iron (M) A126 CL.B (D) B16.4		Cast Iron (M) A126Cl.B (D) B16.1	Steel (M) A234WPB (D) B16.9	Steel (M) A105 (D) B 16.5
.2	Steel (M) A53 (M) A106 (D) B 36.10	Malleable Iron (M) A197 (D) B16.3		Ductile Iron (M) A395 (D) B16.5	Steel (M) A234WPB (D) B16.9	Steel (M) A 105 (D) B16.5
.3	Steel (M) A53 (M) A106 (D) B36.10	Ductile Iron (M) A395 (D) B16.3 (D) B16.39		Ductile Iron (M) A395 (D) B16.5	Steel (M) A234WPB (D) B16.9	Steel (M) A105 (D) B16.5
.4	Steel (M) A53 (M) A106 (D) 36.10	Forged Steel (M) A105 (D) B36.11	Forged Steel (M) A105) (D) B16.11	Steel (M) A216WPB (D) B16.5	Steel (M(A234WPB (D) 16.9	Steel (M) A105 (D) 16.5
.5	Stainless Steel (M) A312 (D) B16.19	Stainless Steel (M) A182 (M) A351 (D) B16.3 (M) A182 (D) B16.11	Stainless Steel (M) A182 (D) B16.11	Stainless Steel (M) A403 (D) B16.5 (D) SP-51	Stainless Steel (M) A403 (D) B16.9 (D) SP-43	Stainless Steel (M) A105 (D) B16.5 Stainless Steel (M) A182 (D) B16.5 (D(Sp-51
.6	Nickel (M) B161 (D) B36.19	Nickel (M) B160 (D) B16.3			Nickel (M) B366 (D) B16.9 (D) SP-43	Steel (M) A105 (D) B16.5 Nickel (M) B160 (D) B16.5 (D) SP-43
.7	Monel (M) B167 (D) B36.19	Monel (M) B164 (D) B16.3			Monel (M) B366 (D) B16.9 (D) SP-43	Steel (M) A105 (D) B16.5 Monel (M) B164 (D) B16.5 (D) SP-43
.8	Titanium (M) B337 Gr2 (D) B36.19	Titanium (M) B367 GrC-2 (D) B16.30 (M) B381GrF-2 (D) B16.11			Titanium (M) B363 (D) B16.9 (D) SP-43	Steel (M) A105 (D) B16.5 Titanium (M) B265 (D) B16.5 (D) SP-43
.9	Aluminum (M) B241 Alloy 606lT6 (D) B36.19	Aluminum (M) B26 Alloy 356F (D) B16.4		Aluminum (M) B26 Alloy 356F (D) B16.1	Aluminum (M) B361 Alloy 606lT6 (D) B16.9	Aluminum (M) B24 Alloy 606lT6 (D) B16.5

Note: (M) Material specifications
  (D) Dimensional specifications

Table 6.6. Suggested Systems for Plastic Lined Piping

System	Plastic Lining	Pipe Material	Type Fittings
1	Saran	Steel	Flanged, Cast Iron/Ductile Iron/Cast Steel
2	Polypropylene (M&D) ASTM F492	Steel	Flanged, Cast Iron/Ductile Iron/Cast Steel/Fabricated Steel
3	PVDF (KYNAR) (M&D) ASTM F491	Steel	Flanged, Ductile Iron/Cast Steel, Fabricated
4	FEP (M&D) ASTM F546	Steel	Flanged, Ductile Iron/ Fabricated Steel
5	PTFE (M&D) ASTM F423	Steel	Flanged, Ductile Iron/ Fabricated Steel
6	PFA	Steel	Flanged, Ductile Iron/ Fabricated Steel

(M) Material specifications
(D) Dimensional specifications

Table 6.7 Suggested Systems for Extruded Plastics

System	Pipe	Screwed	Fittings Socket	Flanged(A)	Flange
1	PVC ASTM D1785	PVC ASTM D2464	PVC ASTM D2466 ASTM D2467	PVC	PVC (M)ASTM D1784 (D) ANSI B16.5
2	CPVC ASTM F441	CPVC ASTM F437	CPVC ASTM F438 ASTM F439	CPVC	CPVC (M)ASTM D1784 (D) ANSI B16.5
3	Polypropylene (M)ASTM D2146 (D) ASTM D1785	Polypropylene (M)ASTM D2146 (D) ASTM D2464	Polypropylene (M)ASTM D2146 (D) ASTM D2467	Polypro-pylene	Polypropylene (M)ASTM D2146 (D) ANSI B16.5
4	KYNAR (M) — (D) ASTM D1785	KYNAR (M) — (D) ASTM D2464	KYNAR (M) — (D) ASTM D2467	KYNAR	KYNAR (M) — (D) ANSI B16.5

Notes: (A) Flanged fittings are fabricated from component parts, elbows, flanges, etc.
(M) Material specifications
(D) Dimensional specifications.

B31.3 — Section 303. Some pipe and fitting manufacturers provide tables of maximum allowable pressures in their literature.

Although the pressure rating of extruded plastic piping can be calculated, the manufacturers of these products publish tables of the pressure ratings of their products. The pressure rating of this category of materials decreases more markedly with increasing temperature and ipe size than

Table 6.8. *Suggested Systems for Fiberglass Reinforced Plastics*

System	Pipe	Screwed	Socket	Fittings Plain End	Flanged	Flanges
1	HL			HL	HL	HL
	NBS			NBS	NBS	NBS
	PS 15-69			PS 15-69	PS 15-69	PS 15-69
2	FW	FW	Molded or		Molded or	Molded or
	ASTM D2996		FW		FW	FW
	ASTM D2997					
3	FW/HL		FW/HL		FW/HL	

Notes: HL: Hand layed up construction
FW: Filament wound construction
FW/HL: Filament wound over hand layed up construction

does metallic piping. Pressure rating tables for filament wound pipe are also provided by the manufacturers in their literature. The pressure ratins of this pipe exhibit a characteristic decrease with increasing temperature and pipe size.

Hand layed up FRP pipe is rated in NBS Product Standard 15-69 (PS 15-69) at 25, 50, 75, 100, 125 and 150 psi. These ratings are applicable to 180°F. The pressure-pipe size-wall thickness table appearing in PS 15-69 is reproduced here in Table 6.11.

FRP Pipe Construction

Epoxy, vinylester or polyester resins are used in the fabrication of filament wound and centrifugally cast FRP pipe. Dimensional standards do not exist for these types of pipe. While the outside diameters of some sizes are the same as NPS the inside diameters differ due to wall thickness differences. For pipe dimensions, manufacturers' catalogs should be consulted.

Polyester and vinylester resins are used in the manufacturer of hand layed up FRP pipe. PS 15-69 in addition to covering dimensional and pressure requirements also specifies the construction of the pipe. While the standard covers pipe sizes 2" through 48", 1" and 1½" pipe can be manufactured.

A variation in the hand layed up FRP products results form the use of hand layed up resin rich liner and corrosion barrier specified in PS 15-69 with a filament wound exterior laminate.

FITTINGS

Screwed and Socket-Weld

Screwed fittings, Figure 6.1. are available in metallic and molded plastic construction. The threads in screwed fittings conform to ANSI B2.1. Socket weld fittings are available in forged steel, stainless steel, molded plastics, molded FRP and filament wound construction.

Table 6.11. Hand Layed Up FRP Pipe Sizes and Pressure Ratings[2]

Material	Specifications	Size Range	Remarks
Carbon Steel	ASTM A53 Type F	through 4″	Continuous welded
	ASTM A53 Type E	4″−26″	Electric resistance welded
	ASTM A53 Type S	through 26″	Seamless
	ASTM A106	through 26″	Seamless
Stainless Steel	ASTM A312	through 12″	Seamless or welded
	ASTM A409	14″−30″	Welded
Nickel 200 Series	ASTM B161	through 8″	Seamless
	(A) ASTM A312	10″ and larger	Welded
Monel 400 Series	ASTM B165	through 8″	Seamless
	(A) ASTM A312	10″ and larger	Welded
Inconel 600 Series	ASTM B167	through 4″	Seamless
Titanium	ASTM B337 Gr2	through 24″	Seamless and welded
Aluminum	ASTM B241	through 24″	Seamless
Extruded Plastics			
Polyvinyl Chloride (PVC)	ASTM D1785	through 12″	140 °F Temperature limit
Chlorinated Polyvinyl Chloride (CPVC)	ASTM F441	through 12″	210 °F Temperature limit
Polypropylene	ASTM D1785 (Dimensions only)	through 6″	160 °F Temperature limit
Polyvinylidene Chloride (PVDF)	ASTM D1785 (Dimensions only)	through 2″	225 °F Temperature limit
Plastics Lined Steel			
Saran	−	1″−8″	175 °F Temperature limit
Polypropylene	ASTM F492	1″−12″	225 °F Temperature limit
PVDF	ASTM F491	1″−10″	275 °F Temperature limit
FEP	ASTM F546	1″−10″	300 °F Temperature limit
PTFE	ASTM F423	1″−10″	500 °F Temperature limit
PFA	−	1″−6″	500 °F Temperature limit
Fiberglass Reinforced Plastic (FRP)			
Custom Contact Molded (hand layed up)	PS 15-69	2″−48″	
Filament wound	ASTM D2996	1″−16″	
Centrifugally Cast	ASTM D2997	1″−12″	
Filament wound over hand layed up	PS 15-69 for hand layed up part	1½″−12″	

(A) Dimensions only.

Table 6.10. Pipe Size and Wall Thickness

Nom. Pipe Size	Out. side Diam.	Sch. 5S	Sch. 10S	Sch. 10	Sch. 20	Sch. 30	Std. Wall	Sch. 40	Sch. 60	Extra Strong	Sch. 80	Sch. 100	Sch. 120	Sch. 140	Sch. 160	XX Strong
½	.840	.065	.083	—	—	—	.109	.109	—	.147	.147	—	—	—	.188	.294
¾	1.050	.065	.083	—	—	—	.113	.113	—	.154	.154	—	—	—	.219	.308
1	1.315	.065	.109	—	—	—	.133	.133	—	.179	.179	—	—	—	.250	.358
1¼	1.660	.065	.109	—	—	—	.140	.140	—	.191	.191	—	—	—	.250	.382
1½	1.900	.065	.109	—	—	—	.145	.145	—	.200	.200	—	—	—	.281	.400
2	2.375	.065	.109	—	—	—	.154	.154	—	.218	.218	—	—	—	.344	.436
2½	2.875	.083	.120	—	—	—	.203	.203	—	.276	.276	—	—	—	.375	.552
3	3.500	.083	.120	—	—	—	.216	.216	—	.300	.300	—	—	—	.438	.600
3½	4.000	.083	.120	—	—	—	.226	.226	—	.318	.318	—	—	—	—	.636
4	4.500	.083	.120	—	—	—	.237	.237	—	.337	.337	—	.438	—	.531	.674
5	5.563	.109	.134	—	—	—	.258	.258	—	.375	.375	—	.500	—	.625	.750
6	6.625	.109	.134	—	—	—	.280	.280	—	.432	.432	—	.562	—	.719	.864
8	8.625	.109	.148	—	.250	.277	.322	.322	.406	.500	.500	.594	.719	.812	.906	.875
10	10.750	.134	.165	—	.250	.307	.365	.365	.500	.500	.594	.719	.844	1.000	1.125	—
12	12.750	.156	.180	—	.250	.330	.375	.406	.562	.500	.688	.844	1.000	1.125	1.312	—
14	14.000	.156	.188	.250	.312	.375	.375	.438	.594	.500	.750	.938	1.094	1.250	1.406	—
16	16.000	.165	.188	.250	.312	.375	.375	.500	.656	.500	.844	1.031	1.219	1.438	1.594	—
18	18.000	.165	.188	.250	.312	.438	.375	.562	.750	.500	.938	1.156	1.375	1.562	1.781	—
20	20.000	.188	.218	.250	.375	.500	.375	.594	.812	.500	1.031	1.281	1.500	1.750	1.969	—
24	24.000	.218	.250	.250	.375	.562	.375	.688	.969	.500	1.219	1.531	1.812	2.062	2.344	—
30	30.000	.250	.312	.312	.500	.625	.375	—	—	.500	—	—	—	—	—	—

Notes:
1. All dimensions are in inches.
2. Schedules 5S and 10S through 12″ are per ANSI B36.19. For sizes 14″ through 30″ are per ASTM A409.
3. Schedules 40S and 80S through 12″ are per ANSI B36.19 and are the same wall thickness as Standard Wall for 40S and Double Extra Strong for 80S.

Table 6.11. Hand Layed Up FRP Pipe Sizes and Pressure Ratings (2)

Pipe Size ID	Minimum Pipe Wall Thicknesses at Pressure Ratings					
	25 psi	50 psi	75 psi	100 psi	125 psi	150 psi
inches	inches	inches	inches	inches	inches	inches
2	3/16	3/16	3/16	3/16	3/16	3/16
3	3/16	3/16	3/16	3/16	1/4	1/4
4	3/16	3/16	3/16	1/4	1/4	1/4
6	3/16	3/16	1/4	1/4	5/16	3/8
8	3/16	1/4	1/4	5/16	3/8	7/16
10	3/16	1/4	5/16	3/8	7/16	1/2
12	3/16	1/4	3/8	7/16	1/2	5/8
14	1/4	5/16	3/8	1/2	5/8	3/4
16	1/4	5/16	7/16	9/16	11/16	
18	1/4	3/8	1/2	5/8	3/4	
20	1/4	3/8	1/2	11/16		
24	1/4	7/16	5/8	13/16		
30	5/16	1/2	3/4			
36	3/8	5/8				
42	3/8	3/4				

The above ratings are acceptable for use to 180°F. For higher temperature service consult the manufacturer.

Table 6.12. Classes 125 and 250 Cast Iron Treaded Fittings (ANSI Standard B16.4)

Temperatures Degrees F	Working Pressures, Nonshock PSIG	
	Class 125	Class 250
−20 to 150	175	400
200	165	370
250	150	340
300	140	310
350	125	280
400		250

Cast Iron

Cast iron screwed fittings are manufactured to the dimensions and pressure ratings specified in ANSI B16.4 through 12″ size. The pressure rating tabulation is shown in Table 6.12 (3).

ASTM A126 Grade B cast iron is predominantly used for these fittings. This cast iron possessing no ductility, when subjected to shock forces, has been known to fracture. Screwed unions are not made in cast iron.

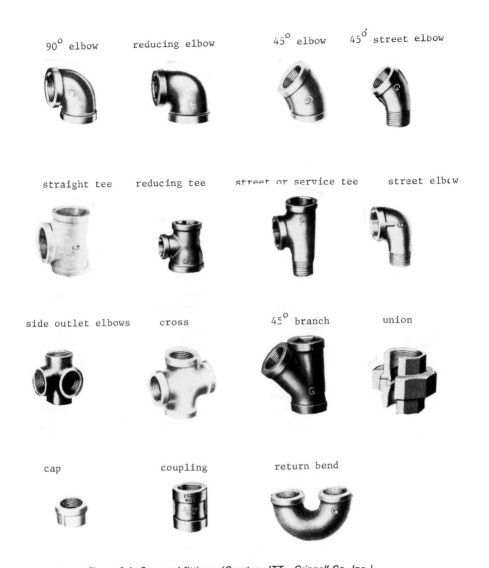

90° elbow   reducing elbow   45° elbow   45° street elbow

straight tee   reducing tee   street or service tee   street elbow

side outlet elbows   cross   45° branch   union

cap   coupling   return bend

Figure 6.1. Screwed fittings. (Courtesy ITT—Grinnell Co. Inc.)

Malleable Iron

Malleable iron screwed fittings are manufactured in accordance with ANSI B16.3 through 6″ size for Class 150 and 3″ for Class 300. Some less used types of fittings are not available in the full size range. The pressure ratings of the fittings are shown in Table 6.13 (3).

Unions are manufactured in accordance with ANSI B16.39.

Table 6.13. Pressure Rating of Malleable Iron Fittings

| | | | Class 300 (PSIG) | |
Temperature Degrees F	Class 150 (PSIG)	Sizes ¼ −1	Sizes 1¼ −2	Sizes 2½ −3
−20 to 150	300	2000	1500	1000
200	265	1785	1350	910
250	225	1575	1200	825
300	185	1360	1050	735
350	150	1150	900	650
400		935	750	560
450		725	600	475
500		510	450	385
550		300	300	300

Threaded Fittings (B16.3)

A line of fittings equivalent to Class 300 but designated AAR (Association of American Railroads) is available and can be used as an alternate to Class 300. AAR fittings include union fittings (elbows, tees) which do not find extensive use in the CPI.

Ductile Iron

Class 300 ductile iron screwed fittings and unions are manufactured to the dimensions of Class 150 malleable iron fittings and unions through the 6″ size. The material specification is ASTM A395. The pressure rating of Class 300 ductile iron fittings is shown in Table 6.14 (3).

Unions are available with iron to iron and iron to bronze seats. Fittings are also offered by one manufacturer in 1000 and 2000 psig ratings.

Forged Carbon Steel

Forged carbon steel screwed and socket weld fittings (Figures 6.1 and 6.2) are manufactured from ASTM A105 carbon steel forgings to the dimensions in ANSI B16.1 through 4″ size. Threaded fittings are designated as 2000, 3000, and 6000 psig and socket weld fittings as 3000, 6000 and 9000 psig in accordance with the procedure outlined in B16.11.

Forged steel unions both screwed and socket weld, are covered by MSS SP-83 but for Class 3000 only. Other pressure classes of unions are available.

The choice of which rated fitting to specify at times depends not so much on the system design pressure but on the need to maximize system integrity.

Table 6.14. Pressure Rating of Ductile Iron Fittings

Temperature Degrees F	Working Pressures, Nonshock PSIG Class 300 Threaded Fittings
−20 to 100	500
150	500
200	480
250	460
300	440
350	420
400	400
450	380
500	360
550	340
600	320
650	300

[1]Brass to iron seat unions have a maximum temperature of 450°
in accordance with the ASME Boiler Code ratings on brass seat
materials.

Stainless Steel

Stainless steel threaded fittings are rated at 150 (300 cwp), 2000, 3000 and 4000 psig and socket weld fittings at 3000 and 6000 psig. Unions are rated at the same as fittings. The 150 psig fittings are made through the 4″ size. The higher pressure fittings are made through 1 ½ ″, 2 and 4″ sizes depending upon the type of fitting.

Dimensions of 150 psig stainless conform to ANSI B16.3, Class 150 Malleable Iron while the higher pressure fittings are made to the dimensions of ANSI B16.11, Forged Steel Fittings.

The 150 psig fitting are manufactured from ASTM A182 (the stainless steel grades) or ASTM A351 depending on the type fitting made. The higher pressure fittings are manufactured from ASTM A182 material.

Non Ferrous Materials

Nickel threaded fittings are rated at 150 psig (300 cwp) and are made to the dimensions of ANSI B16.3, Class 150 Malleable Iron. Nickel material conforming to ASTM B160 and cast nickel, are used for the manufacture of these fittings through the 2″ size. High pressure threaded fittings rated at 2000, 3000 and 6000 psig and socket weld fittings are made to B16.11 dimensions.

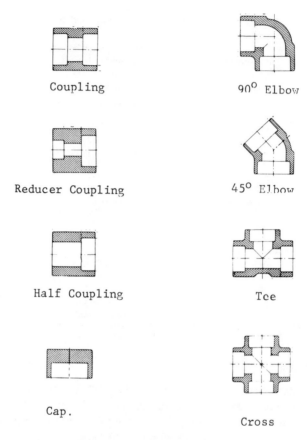

Coupling

90° Elbow

Reducer Coupling

45° Elbow

Half Coupling

Tee

Cap.

Cross

Figure 6.2. Metallic socket fittings.

Monel threaded fittings are rated at 150 psig (300 cwp) and are made to the dimensions of B16.3, Class 150 Malleable Iron through the 2″ size. Monel material conforming to ASTM B164 and cast monel are used in the manufacture of these fittings.

Titanium threaded fittings are made through the 2″ size. The fittings are rated at 150, 2000 and 3000 psig. The 150 psig fittings meet the dimensional requirement of ANSI B16.3, Class 150 Malleable Iron and are manufactured from cast titanium, ASTM B367 Grade C-2. The 2000 and 3000 psig fittings meet the dimensional requirements of ANSI B16.11 and are manufactured from titanium forgings, ASTM B381 Grade F-2. Unions and threaded couplings are machined from titanium bar stock, ASTM B348, Grade 2, for all ratings.

Aluminum threaded fittings are made through the 6″ size. The fittings are rated at 150 psig to 400F. The fittings meet the dimensional requirements of Class 125 cast iron threaded fittings, ANSI B16.4. Cast aluminum, ASTM B26, Alloy 356F is one of the materials out of which these fittings can be made. Couplings and nipples are made from another aluminum alloy, ASTM B241, Alloy 6060-T6. Also available is a line of traced aluminum pipe fittings and flanges in which the tracer is an integral part of the pipe, fitting or flange.

Molded Thermoplastics

Thermoplastic fittings and unions are not rated in the same manner as are metallic fittings. Thermoplastic fittings and unions are rated the same as the thermoplastic pipe to which they are attached. A Schedule 40 socket fitting would be rated the same as Schedule 40 pipe, a Schedule 80 threaded fitting the same as Schedule 80 threaded pipe. Schedule 40 pipe should not be threaded as insufficient wall thickness remains for pressure service. Manufacturers catalogs are a source of pressure capability information.

The Table 6.15 indicates fitting ratings, material and dimension specifications and size availability limits.

FRP (Fiberglass Reinforced Plastics)

Socket type fittings are available in molded or filament wound construction using epoxies, vinylester or polyester resins. There are no dimensional standards that apply to the socket type fittings. These fittings are made in the size range of 1″ through 16″.

No two socket designs are the same. Each manufacturer has incorporated some feature into the socket design which creates a difference. The difference may be the degree of taper in the fitting to match the taper on the pipe or no taper at all on the pipe.

There is no standard pressure rating system which covers all of these fittings as there is for metallic fittings. Each manufacturer rates their fittings at somewhat different pressures than other manufacturers. To achieve the full rating of any system it is not advisable to intermix the pipe and fitting of the various manufacturers in any one system.

Buttwelding Fittings

Buttwelding metallic fittings are made to the following specifications.
ANSI B16.9 Wrought Steel Buttwelding Fittings.
MSS SP-43 Wrought Stainless Steel Butt-Welding Fittings.
MSS SP-43 covers fittings in Schedule 5S and 10S wall thickness only. The fittings made to this standard, which can also be applied to non-ferrous materials are primarily for corrosion resistant service. Their pressure rating as designated in the standard is less than that for the pipe to which the fittings are

Table 6.15. Specifications and Limitations of Molded Thermoplastic Fittings

Rating	Material and Dimension Specification	Size Limit
PVC (polyvinyl chloride)		
Threaded		
Sch. 80	ASTM 2464	4″
Socket*		
Sch. 40	ASTM 2466	8″
Sch. 80	ASTM 2467	8″
*Larger sizes available in fabricated fittings.		
Threaded		
Sch. 80	ASTM F437	4″
Socket		
Sch. 80	ASTM F439	6″
PP (polypropylene)		
Threaded**		
Sch. 80	(M)ASTM D2146 Type 1	4″
Socket		
Sch. 40	(M)ASTM D2146 Type 1	8″
Sch. 80	(M)ASTM D2146 Type 1	6″
**Threaded systems not recommended for pressure service.		
PVDF (Kynar)***		
Threaded		
Sch. 80	No specifications	2″
Socket		
Sch. 80	No specifications	2″
***PVDF are manufactured to the same general dimensions as other molded thermoplastic fittings.		

(M) Material specifications. (D) Dimensional specifications.

attached. For stainless steel and non-ferrous fittings rated the same as the pipe the fitting would be manufactured in accordance with ANSI B16.9.

With two exceptions thermoplastic buttwelding fittings are not available. The exceptions are polyethylene and polybutadiene. Buttwelding fittings for these materials are used with the same piping material for purposes such as gravity and pressure waste lines to very large sizes.

Thermosetting plastic butt fittings are made to NBS Standard PS 15-69. These fittings are rated in the same manner as the pipe, 25 psig, 50 psig, etc. The fittings are attached to the pipe using the butt and strap method outlined in Joining Systems section.

Elbow
(Can be long or
short radius)

Reducer
Can be concentric or eccentric

Cap

Tee
(Can be straight or reducing)

180° Return
(Can be short or long radius)

Lap Joint Stub End

Cross

Straight 45° Lateral

Saddle

Figure 6.3. Metallic butt welding fittings. Courtesy ITT-Grinnell Co., Inc.

FLANGES AND FLANGED FITTINGS

Flange Types

Metallics — The types of metallic flanges available are shown in Figure 6.4. Dimensions and pressure ratings of steel, low alloy and stainless steel flanges are covered by ANSI B16.5. Non-ferrous flanges are generally made to the same dimensions and are listed in B16.31. Cast iron flanges are made to the dimensions shown in ANSI B16.1. Class 150 ductile iron flanges are made to the outside diameter and bolt circle dimensions of Class 125 cast iron flanges and Class 150 steel flanges, which are the same.

Figure 6.4. Metallic flanges. Courtesy ITT-Grinnell Co., Inc.

In addition to the above flanges there are several lines not covered by the referenced standards. One is MSS SP-51, 150 LP Corrosion Resistant Cast Flanges and Flanged Fittings. The primary difference is that the flange thickness is less than called out in ANSI B 1.5. Threaded and blind flanges only are covered by MSS SP-51.

ANSI B16.5 includes flanges through 24". Above 24" a series of lightweight steel flanges is manufactured. These flanges are made to the outside diameter and bolt circle dimensions of cast iron flanges (ANSI B16.1). For specifics on other dimensions, gaskets and bolting requirements the manufacturer's catalogs should be consulted.

Flange Facings — The type of flange facings available are shown in Figure 6.5. The flat and raised facings are the most commonly used. The flat facings are found on the following flanges:

    ANSI B16.1   Class 125 Cast Iron
    MSS SP-51   Stainless Steel
    NBS PS 15-69   Hand Layed Up Flanges

Raised facings can be found on these flanges:

    ANSI B16.1   Class 250 Cast Iron
    ANSI B16.5   Steel Pipe Flanges and Flanged Fittings

The other facings shown for metallic flanges are used to a lesser degree and in situations where an extraordinary flange seal is required.

Molded plastic and hand layed up flanges while nominally being classified as flat face may be provided with a facing to grip the flange gasket.

The flat gasket contact surface of metallic flanges can be either smooth or serrated in accordance with MSS SP-6. Raised face flanges usually will be provided with a serrated surface while flat face flanges will have a relatively smooth face. One exception applies to stainless steel flanges manufactured to MSS SP-51 which although having flat faces are provided with a serrated finish.

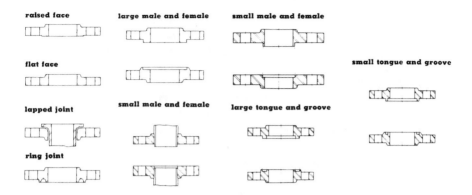

Figure 6.5. Flange facings. Courtesy ITT-Grinnell Co., Inc.

Plastics — Plastic flanges, whether reinforced or not reinforced (with fiberglass) generally conform to the outside diameter and always to the bolt circle dimenisons of ANSI B16.1 (Class 125 cast iron) and B16.5 (Class 150 steel) flanges. For other dimensions the manufacturer's catalog should be consulted except for the minimum thickness of hand layed up FRP flanges which is listed in NBS PS15-69.

Unreinforced plastic (PVC, CPVC, polypropylene, etc.) flanges are made in the threaded, socket and blind types. Reinforced plastic (press molded, hand layed up FRP) flanges are made in the socket, stub end, lap joint and blind types. Not all manufacturers will necessarily produce all the type flanges listed above.

Flange Ratings

Metallic — Metallic flanges are rated according to a class basis. Each class has a pressure-temperature relationship. For steel, alloy and non-ferrous flanges classes 150, 300, 400, 600, 900, 1500 and 2500 apply. The pressure-temperature relationship for each of the classes is shown in ANSI B16.5 for steel and alloy flanges and ANSI B16.31 for non-ferrous flanges.

Cast iron flanges are rated as class 125 and 250. Classes 25 and 800 are also shown in ANSI B16.1 but are not used very often.

Ductile iron flanges are rated as Class 150. Their pressure-temperature relationship is the same as that for Class 150 carbon steel flanges.

Plastic — Unreinforced thermoplastic flanges are usually rated at 150 psig at 73°F. The ratings decline with increasing temperature according to the characteristic of each material.

Considering the pressure rating of press molded, hand layed up and filament wound FRP flanges, a rating standard exists only for the hand layed up flange which is the same as for hand layed up FRP pipe. Since press molded and filament wound flanges are not made to a rating standard, the pressures each are capable of withstanding must be determined from each manufacturer's catalog. In general, these flanges span the range 100 psig to 300 psig.

Flanged Fittings — Flanged fittings can be of metallic (Figure 6.6) or plastic construction. The metallic flanged fittings are either of cast construction or fabricated of welding together fittings and flanges. One type of fitting is fabricated from bent pipe for elbows or welded pipe for tees with machine formed laps on the fitting ends. Backing up these end laps are lap joint type flanges which are free to rotate on the pipe, hence the fittings can be called "loose flange fittings." Presently these loose flange fittings are commercially available in sizes 1″ through 4″ size range.

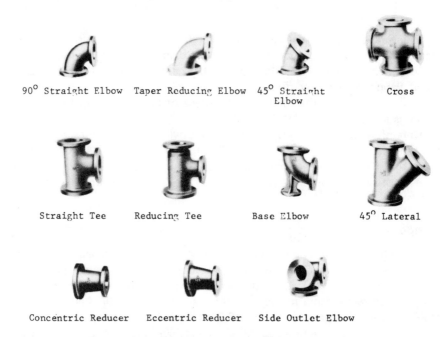

| 90° Straight Elbow | Taper Reducing Elbow | 45° Straight Elbow | Cross |

| Straight Tee | Reducing Tee | Base Elbow | 45° Lateral |

| Concentric Reducer | Eccentric Reducer | Side Outlet Elbow |

Figure 6.6. Flanged fittings. Courtesy ITT-Grinnell Co., Inc.

Plastic — Plastic flanged fitting can be made either from component parts, as molded piece or filament wound construction. PVC, CPVC, polypropylene and PVDF flanged fittings are fabricated from components. Thermosetting plastic (epoxies, vinylester, polyesters) are constructed as one piece press molded filament wound or from components dependent on the manufacturer's product line. Custom molded (hand layed up) thermosets are fabricated from component parts which are made in accordance with NBS

Table 6.16. Metallic Flanged Fittings

Cast Material	Material Spec.	Dimension & Press. Spec.	Pressure Rating	Size Range
Cast Iron	ASTM A126 Class B	ANSI B16.1	Class 125 Class 250	1″ − 48″
Ductile Iron	ASTM A395	ANSI B16.1 (Class 125)	Class 150	½″−12″
Steel	ASTM A234 WCB	ANSI B16.5	Class 150-2500	½″−24″
Stainless Steel	ASTM A351 Series	MSS SP-51	230 psi @ 100F 150 psi @ 500F	¼″−12″
Aluminum	ASTM B26 Alloy 356-F	ANSI B16.1 Class 125	150 psi @ 100F to 400F	½″−6″

PS15-69. The fitting pressure rating for all the plastics is the flange rating.

JOINING SYSTEMS

Pipe joining systems (4) can be classified as:
Screwed
Socket
Buttwelded
Flanged
Mechanically Coupled
and apply to both metallic and plastic piping.

SCREWED SYSTEMS

Metallic Systems — As a rule of thumb it is suggested that screwed piping be limited to the 2″ size for utilities services and 1½″ for process services. If stainless steel or non-ferrous piping is involved, the maximum threaded size may be limited to 1″. Pipes thinner than Sch. 40 or 40s should not be threaded. As the pipe sizes becomes larger the difficulty of threading the pipe increases as does the assembly. Other methods of assembly then become more economic. The size availability of screwed fittings also influences the maximum size of screwed construction depending on the material of construction of the fitting.

Thermoplastic — The same size limitations as suggested for threaded metallic pipe can apply to thermoplastic pipe, 2″ for utilities and 1½″ for process services. Pipe thinner than Sch. 80 should not be threaded. A great deal more care must be taken when cutting threads in plastic pipe as the material is relatively soft. The manufacturer's recommendations should be followed for cutting threads. When assembling a threaded plastic system the use of a strap wrench is suggested as the use of a pipe wrench would probably damage the pipe. PTFE tape or PTFE paste type compounds can be used as thread compounds for screwed plastic pipe.

Thread Compounds (Pipe Dope) — The use of lubricant is necessary when assembling threaded joints without the use of excessive force to obtain full thread engagement with the ANSI B2.1 thread system. To insure a tight threaded joint, a compound is added to the lubricant which will act as a filler to compensate for imperfect threads and reduce the possibility of leakage through the joint. This combination of materials is termed a thread compound or pipe dope.

There are many thread compounds available today. Approximately 60 companies market this material. In the past a mixture of white lead and oil was used extensively. A mixture of letharge and glycerine still has limited use where a very tight seal is required. Another compound which expands upon setting up finds use in refrigerator services.

More recent compounds make use of a dispersing of PTFE in an organic or inorganic base. A PTFE tape is also used as a thread compound. With the variety of compounds available, the user should satisfy himself that the compound is suitable for the intended service.

Application of the thread compound is critical. The compound should not be applied to the threads in such a manner that the compound enters the pipe. Always apply the compound behind the first thread at the end of the pipe or male thread of a fitting. If the compound enters the pipe it could react with the fluid being conveyed or could block small openings as in instruments.

Socket System

Metallic — Socket type metallic systems are used in situations where as leak-free construction as possible is required in piping systems to the 1½″ size. For 2″ systems and larger, buttwelded construction serves the same purpose. This is not to say that screwed systems are inherently leak prone. But circumstances do exist, such as cyclic services or workmanship problems, where a socket fitting welded into the line provides greater system integrity.

Installation of metallic socket fitting requires a difficult welding technique. Another aspect of the use of these fittings which causes concern is the space that is created between the end of the pipe and the shoulder in the fitting. Proper installation calls for backing the pipe out of the fitting 1/16″ before welding. Corrosion could start in this area as a result of materials lodging there.

Even with the factors that tend to increase the cost of the socket system and reduce its utility, where a high level of system integrity is required, the metallic socket system serves its purpose within the suggested size range.

Plastics

Thermoplastics — Socket construction in thermoplastic systems is an alternate to screwed construction through 4″ and one of the primary joining systems for sizes larger than 4". This construction has the advantage of higher pressure capabilities than screwed construction.

There are two methods of installing the pipe in the socket of the fitting (or flange). One is with a solvent cement which is used with PVC and CPVC materials. The other method utilizes a process called thermal fusion (also called thermosealing or thermalbonding). This latter process applies to polypropylene and PVDF which cannot be cemented. In both of these methods the outside surface of the pipe and inside surface of the socket are prepared in such a manner that when joined, adhesion occurs.

The solvent cement is a solution of finely divided resin in a solvent such as tetrahydro furan. The cement will adhere to the surface of the pipe and socket. When the joint is made up the cement softened surfaces come in contact displacing some material in the joint. The joint "cures" through the evaporation of the solvent from the cement. Manufacturers of the pipe, socket fittings and cement, make instructions available on the procedure to follow for making this type of joint.

The thermal fusion technique involves heating the pipe outside surface and fitting inside surface till they are soft. When the pipe is inserted into the fitting there is a displacement of material due to the interference fit designed into the socket inside surface. When cooled, the joint is ready for use. Here again manufacturers instructions should be followed.

A variation of the fusion technique involves a socket fitting with resistance wire in the wall of the socket. After the pipe is inserted into the socket current is applied to the wire fusing the pipe into the socket. This method is used on polypropylene drain systems.

The significant differences between the screwed and socket systems in thermoplastic materials is that the pressure capabilities of socket system are greater than those of screwed systems but socket joints cannot be dismantled as can screwed joints.

Fiberglass Reinforced Plastics (FRP) — Socket joints are used extensively in FRP systems particularly with filament wound and centrifugally cast pipe systems. Socket designs do not conform to any standards as they do for thermoplastic systems. The socket design of the fittings made by one firm is different than that of any other manufacturer's design. Joint systems include tapered socket, straight socket with full pipe outside diameter and straight socket with reduced pipe outside diameter. The joints are assembled using an adhesive specific for the piping material involved. Each manufacturer's instruction should be consulted for the procedures to follow in making thermosetting (FRP) socket joints.

Butt Welding

Metallic — Joining pipe to pipe, pipe to fitting, flange to pipe or fitting with a buttweld (also called circumferencial or girth weld) is the most common method of joining metallic systems in sizes 2″ and larger. This method is also used to joint smaller sizes but there are other methods better suited which have been previously cited.

In making buttwelds items such as the welding process, filler material (weld rod or wire), pipe or fitting end design and preparation, weld examination and inspection are taken into consideration.

The use of backing rings is argumentative. The rings are inserted at the inside wall of the joint to be welded. Their purpose is to prevent weld material from protruding into the interior of the pipe. The argument against their use is that crevices are formed between the outside surface of the ring and inside wall of the pipe where corrosion could be initiated.

Plastic

Thermoplastics — Only two thermoplastic piping materials are usually joined by buttwelding. These are polyethylene and polybutadiene. The end of the pipes are cut square, placed in a hydraulic or manually operated rig face to face, where they are placed against a heated plate. When the ends have been heated to the point where they are soft and pliable the heated plate is removed and the pipe ends forced into each other. The pipe is held in this position till the joint has cooled at which time the pipe is ready to be installed.

Fiberglass Reinforced Plastics (FRP) — Thermosetting piping, the hand layed up variety is not buttwelded in the actual sense. The ends of the pipe are butted together and then the joint is overwrapped with successively wider layers of resin impregnated fibeglass till a wrap as thick as the pipe wall has been constructed. When this wrap is cured it is as strong as the pipe. This pipe joining method goes by the name "Butt and Strap."

Flanged Joints

Flanges can be used in almost any type of piping system where either attachment to flanged equipment is required or there are no other means of taking the piping apart (as in socket or butt welded construction). Flanged attachment to the piping can be made with screwed, socket, butt welded, stub end, insert and butt and strap joining systems.

Stub End Joints — The stub end system makes use of the lap joint flange. An example of this configuration is shown in Figure 6.5. The material of construction of the flange is usually different from that of the stub end. This combination finds its greatest use in stainless steel and non-ferrous piping where the stub end is the same as the piping material but the flange can be carbon steel or ductile iron. This is a cost saving over the flange the same as the pipe material.

Another advantage of lap joint type flange is that the flange is free to rotate

218

on the stub end. When a flange is welded or otherwise firmly attached to the pipe the vertical centerline of the pipe locates the bolt holes on the flanges so when flanged end systems are bolted together the bolt holes are always oriented in the same manner. But the lap joint flange being free to rotate or loose on the stub end does not require this orientation reducing fabrication and installation costs.

One variation of the stub end-lap joint flange combination involves forming a lap on the pipe end rather than welding a stub end to the pipe. This operation is also called cold flanging. It finds application not only in stainless steel and non-ferrous but carbon steel piping. But in order for the carbon steel piping to be cold flanged it must be more ductile than A53 or A106 pipe. ASTM A587 carbon steel pipe was developed with a 40% ductibility for cold flanging and bending. This pipe is presently available in the sizes 1″ through 4″ with the possibility of being available through 8″ in the future.

The one drawback to the cold flanged piping system is that a piece of expensive machinery (up to $30,000) is required to form the laps.

Another variation utilizes an insert into which the pipe or fitting is roll expanded into the end of the insert. A flange made to ANSI B16.5 outside diameter and bolt circle dimensions fits over the pipe or fitting end and rests against a shoulder on the insert. The insert is the same material as pipe or fitting while the flange can be carbon steel. The insert has to be installed in a straight length pipe. The fittings for this system are not made to ANSI B16.9 dimensions. Each fitting end is designed with a straight section called a tengent for the installation of the insert. The fittings can also be welded to pipe if flanged joints are not needed.

While the insert system is applicable only to metallic piping, the stub end system has a counterpart in thermosetting (FRP) system. One FRP system has a stub end fitting with which is used the equivalent of the lap joint flange in FRP construction. Steel plate flanges are used in other FRP stub systems.

Flange Facing Problems — When joining flat face to flat face or raised face to raised face metallic flanges no problems are encountered. But when joining a raised face flange to a flat face flange extra care should be exercised to avoid damaging one of the flanges. This situation can be found when joining a Class 125 cast iron flange to a Class 150 raised face steel flange. It is possible to cause the cast iron to fail if excessive pressure (i.e. more than is required to seal the joint) is applied when bolting the flanges together. A similar situation is possible when bolting a plastic flange (usually flat face) to raised face flange.

To avoid this problem remove the raised face from one of the flanges or place a spacer around the raised face to produce a flat face. The third approach is to use care in tightening the flange bolts to avoid flange failure but still effecting a seal.

Gaskets — Gaskets are used in flanged joints to effect a seal between the two flange faces. Gasket materials vary from compressed asbestos (asbestos with synthetic rubber binder) to synthetic elastomers (Neoprene, SBR, Buna-N, EPDM, CPE, etc.), to PTFE (solid or envelope type), to spiral wound and various other metallic gasket forms. Table 6.17 shows the various type of gaskets that can be used in flanged joints. Gaskets are selected on the basis of previous service or analytical approaches (5).

Compressed asbestos gaskets have been the most widely used gasket for metallic piping systems with the 1/16″ and 1/8″ thickness predominating. Non-asbestos materials are now making appearance in the market. The gasket forms used are full face for flat face flanges, ring type for raised face and tongue and groove flanges. The maximum service temperature for this material is 750°F. Dimensional standards for nonmetallic flange gaskets are shown in ANSI B16.21.

Synthetic elastomers are used as gasket material extensively in plastic systems. Elastomers are particularly suited to this service as plastic flanges cannot sustain high bolt loading and elastomers require a low compressive force as compared to other materials to effect a seal. The minimum gasket thickness should be 1/8″. Full face gaskets predominate as plastic flanges are mostly flat face.

PTFE (solid and envelope type) requiring a relatively high compressive force to seal are used more often with metallic than plastic flanges. PTFE envelope type gaskets are available with asbestos or Neoprene inserts. With plastic flanges, the Neoprene insert should be used rather than the asbestos insert. The envelope gasket with the Neoprene insert cannot be used at as high a temperature (250°−300°F) as can the solid PTFE gasket or the envelope with the asbestos insert (500°F). When used with plastic flanges, this temperature limitation presents no problem as plastic systems are not used generally at temperatures above the former range.

Filled metallic gaskets requiring a higher compressive force to seal, need a smaller contact area, as is found on raised face or small tongue and groove flanges, to produce the necessary compressive force. Asbestos filled metallic gaskets can be used to temperatures to 1000°F. If the filler material is PTFE as can be used in spiral wound gaskets, the temperature would be limited by the PTFE to about 500°F. In addition to alternate filler materials, the metal part of the spiral in these gaskets is available in a variety of metals e.g., carbon steel, stainless steel, nickel, etc. Use in flange sizes through 24″ and pressure ratings through Class 2500 is common.

Metallic rings serve as gaskets in ring type flanges. Dimensions of the rings and flange grooves in which they fit are covered by ANSI B16.20. The groove dimensions are also shown in ANSI B16.5. The rings are made of soft iron, low carbon steel, chrome moly steel, a number of stainless steels, Monel and nickel etc. This gasket system finds use in high pressure and/or high temperature service particularly in the petroleum industry.

Table 6.17. Gaskets: Materials and Configuration.

Gasket material			Sketches
Rubber without fabric or a high percentage of asbestos fiber: Below 75 Shore Durometer 75 or higher Shore Durometer			
Asbestos with a suitable binder for the operating conditions		⅛ thick ¹⁄₁₆ thick ¹⁄₃₂ thick	
Rubber with cotton fabric insertion			
Rubber with asbestos fabric insertion, with or without wire reinforcement		3-ply 2-ply 1-ply	
Vegetable fiber			
Spiral-wound metal, asbestos filled		Carbon Stainless	
Serrated steel	Asbestos filled		
Corrugated metal, asbestos inserted or Corrugated metal, jacketed asbestos filled	Soft aluminum Soft copper or brass Iron or soft steel Monel or 4-6% chrome Stainless steels		
Corrugated metal	Soft aluminum Soft copper or brass Iron or soft steel Monel or 4-6% chrome Stainless steels		
Flat metal jacketed asbestos filled	Soft aluminum Soft copper or brass Iron or soft steel Monel 4-6% chrome Stainless steels		
Grooved iron or soft steel with or without metal jacket	Soft aluminum Soft copper or brass Iron or soft steel Monel or 4-6% chrome Stainless steels		
Solid flat metal	Soft aluminum Soft copper or brass Iron or soft steel Monel or 4-6% chrome Stainless steels		
Ring joint	Iron or soft steel Monel or 4-6% chrome Stainless steels		

Bolting — Of the bolting materials available, three appear to be used more often than any others. The first is ASTM A307 which is a carbon steel material suitable for temperatures to 450°F. For higher temperatures and situations where greater compressive force is required, the use of ASTM A193 Grade B7 studs is common. For low temperature service ASTM A320 Grade L7 studs are used. Corrosion resistant bolting of 304 and 316 stainless steel is also available. The Table 6.18 indicates possible bolt-nut or stud-nut combinations.

Table 6.18. Nuts & Bolts

Bolt	Bolt Form	Nut[a,c]
A307 Grade A	Hex Head[b]	A307 Grade A
A307 Grade B	Heavy Hex Head[b]	A307 Grade B
A193 Grade B7	Stud	A194 Grade 2H
A320 Grade L7	Stud	A194 Grade 7
A193 Grade B8(304)	Stud	A194 Grade 8(304)
A193 Grade B8M(316)	Stud	A194 Grade 8M(316)

[a]A307 Grade A nut is a regular hex series. Other nuts are heavy hex series.
[b]Hex head bolt dimensions are specified in ANSI B18.2.1.
[c]Hex nut dimensions are specified in ANSI B18.2.2.

ASTM A307 Grade A and B bolting systems differ in two aspects. While both exhibit a minimum tensile strength of 60,000 psi only Grade B meets a maximum tensile strength requirement of 100,000 psi. Only Grade B is listed in several codes (ASME Unfired Pressure Vessels and ANSI Pressure Piping Codes). The A307 Grade A can be used anywhere where the A307 Grade B system can be used, except in cases where the codes apply or their use is governed by company standards.

When assembling a bolting system it is advisable to apply a lubricant to the bolt or stud threads to reduce the torque necessary to tighten the system and permit easier disassembly.

Mechanically Coupled Joints (6)

A segmented mechanical coupling (Figure 6.7) is available in size 3/4″ through 24″. The coupling can be used with standard weight pipe, through wall pipe (Sch. 5S, 10S) and cast iron or ductile iron underground pipes. The pressure rating of the coupling can vary from 300 psig to 1000 psig depending on size and material of construction (malleable iron, aluminum, etc.). The grooves in the pipe into which the coupling end fits, are either cut into the pipe wall or rolled into the wall for thin wall pipe. The gasket which is the sealing member fits over the pipe ends and is contained by the coupling. The gaskets are made from a variety of synthetic elastomers (Buna N, EPDN, Neoprene, etc.) Fittings, butterfly and check valves are available for use with grooved end pipe. The fittings are made of malleable or ductile iron and are also available in steel, aluminum, stainless steel and cast iron (for underground piping). The butterfly valves are elastomer lined. The check valves incorporate the dual disc design with an elastomer seal on the seating surface.

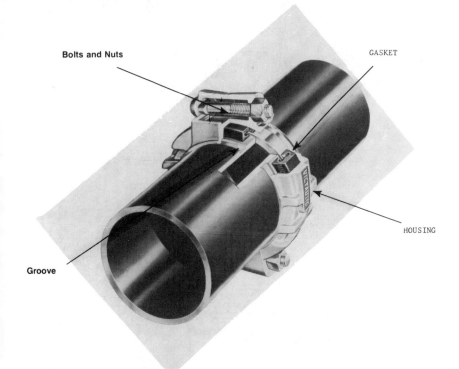

Bolts and Nuts

GASKET

HOUSING

Groove

Figure 6.7. Segmented mechanical coupling. Courtesy Victaulic Company of America.

Another mechanical coupling uses a cylindrical barrel with elastomer seals at each end. The seals are activated by tightening the through bolts which in turn force a collar at each end of the coupling into the elastomer causing it to seal the pipe outside diameter to the coupling inside diameter. This coupling sometimes known as a Dresser type coupling, can be used in both underground and aboveground utility services (Figure 6.8).

A third type of mechanical coupling (Figure 6.9) can be used in corrosive service. The sealing member is PTFE seals on the pipe outside diameter and ends. The seal and coupling body are held in place with a wedging action of collars produced by tightening the coupling. In the 1/4″ through 1″ size the coupling construction resembles a pipe union. The 1½″ through 4″ size incorporates a bolted design to tighten the collars. In addition to pipe couplings,

223

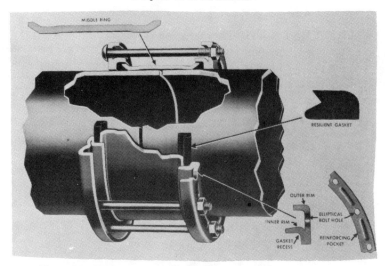

Figure 6.8. Bolted through coupling. Courtesy Dresser Industries.

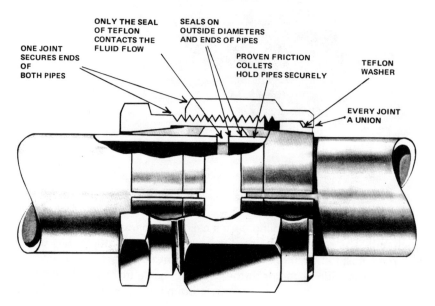

Figure 6.9. Mechanical coupling for pipe and tubing. Courtesy Stewart Controls Co.

other fittings and ball valves are also made with this coupling end design. Stainless steel is used where fluid contacts the fitting or valve body. The components are pressure rated from 500 psig non shock for the smaller sizes to 125 psig non shock for the larger sizes.

VALVES (7,8,9,10,11)

Valves in a piping system control the flow of the fluid in the system. This control is in the form of blocking (off-on throttling rate of flow control) or checking (directional control of flow) action.

Valve Types and Function

> Gate — Blocking
> Globe — Blocking/Throttling
> Butterfly — Blocking/Throttling
> Ball — Blocking/Throttling
> Plug — Blocking/Throttling
> Diaphragm — Blocking/Throttling
> Pinch — Blocking/Throttling
> Check — Checking

Most of the valve types shown can serve the double purpose of blocking and throttling. Each type valve has its own throttling and pressure drop characteristic.

Valve Pressure Ratings and Standards

Valve pressure ratings are dependent on the valve type, material of construction and whether the valve is threaded or flanged. Following is a list of standards dealing with pressure rating and dimensions of some valves.

ANSI	B16.10	Face to face and end to end dimensions of ferrous valve.
	B16.34	Steel valves, flanged and butt welded end.
MSS	SP-42	150 LP corrosion resistant cast flanged valves
	SP-67	Butterfly valves
	SP-70	Cast iron gate valves, flanged and threaded ends
	SP-71	Cast iron swing check valves, flanged and threaded ends
	SP-72	Ball valves with flanges or butt weld end for general service
	SP-78	Cast iron plug valves
	SP-80	Bronze gate, globe, angle and check valves
	SP-81	Stainless steel, bonnetless, flanged, wafer, knife gate valves
	SP-84	Steel valves — socket welded and threaded ends
	SP-85	Cast iron globe and angle valves, flanged and threaded ends.

Gate Valves (12)

Gate valves (Figure 6.10 A,B&C) are used for blocking service, to turn the flow of fluid off or on. They can be manufactured from any of the conventionally used materials and pressure rated from 125 psig to as high as 2500

225

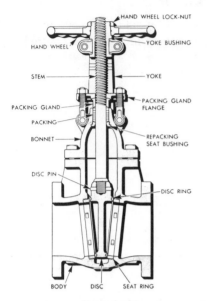

Figure 6.10A.
Solid wedge, taper seat, rising stem,
bolted bonnet, outside screw & yoke.

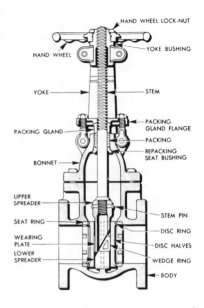

Figure 6.10B.
Double disc parallel seat, rising stem,
bolted bonnet, outside screw & yoke.

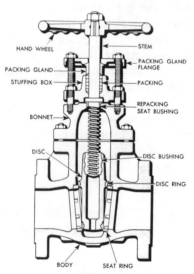

Figure 6.10C.
Solid wedge, taper seat, non rising stem,
inside screw.

Figure 6.10. Courtesy Stockham Valves and Fittings.

psig. This type of valve exhibits a low pressure drop due to its straight through flow pattern.

Bonnets — The valve consists of five basic parts, the body, bonnet, stem, disc and seats. On threaded valves the bonnet to body joint can be a threaded-in or -over arrangement, union or bolted attachment. For most flanged gate valves a bolted bonnet is usually employed (Figure 6.10A).

Screw bonnet valves are not necessarily considered to be repairable and are obtainable at a lower cost than the union or bolted bonnet valves. The union and bolted bonnet valves can be rated at a higher pressure than the screwed bonnet valve.

Stems — The valve stem designs can be classified as rising or non-rising. The non-rising stem will have threads on that portion of the stem which is inside the bonnet which will also be threaded (Figure 6.10C). The rising stem can either be threaded on that portion inside the bonnet or above the bonnet. The former is termed "rising stem — inside screw" and the latter OS & Y (outside stem and yoke). (Figure 6.10A).

The use of the non-rising stem permits valve installation without any allowance for the stem movement beyond its fixed position. Also, the non-rising stem affords the possibility of better containment of the fluid in the valve as stem rotates in the packing rather than moving through it. On the other hand, the stem threads are always being subjected to the fluid in the valve. If the threads corrode the valve would be difficult, if not impossible to operate. Materials of construction selection has significance for other than general service life of the valve. Another consideration when using the non-rising stem is that fluids containing solids can impair the operation of the stem due to the solids jambing the stem threads.

Inside screw rising stems are similar to non-rising stems as to the problems that can be encountered. Their advantage is a lower cost than the OS & Y in the smaller valves, but repairability of these valves is limited. The OS & Y type covers the entire range of valve sizes and commonly for sizes above 2″. The OS & Y construction permits more force to be applied to the valve stuffing box enabling the valves to be designed for higher pressure. Also this stem type includes a "back seating feature" to permit packing replacement while the valve is in service. A disadvantage of the OS & Y stem is that the threads are exposed to the atmosphere and are subject to corrosive effects that may result.

Stem packing materials that can be used vary widely. Teflon impregnated asbestos has found extensive use.

Discs — Three types of discs are used in gate valves, solid wedge, (Figure 6.10A, 6.10C) split wedge, (double disc) (Figure 6.10B) and flexible wedge. The seating surfaces are usually constructed at an angle to each other, hence, the wedge shaped cross section to the closure discs. The split and flexible wedges overcome some of the problems of valve closure caused by seat wear and temperature changes. As the valve size increases the tendency is to use split or flexible wedge instead of the solid wedge. One further consideration involves the wedge to be specified when handling fluids containing solids. In this type of situation the solid wedge is preferred over the split or flexible wedges as the possibility exists that solids will lodge between the disc causing valve closure to be difficult or in the extreme, not functioning at all.

Seats — To avoid excessive valve seat wear the mating surfaces can be hardened or Stellite faced to reduce the wear. With one seating surface in the valve body and the other on the wedge, one approach is to provide the body seat with the harder surface so that the wear will be primarily on the disc which can be replaced.

The valve body seats can either be integral with the body or can be inserts that are threaded or welded into place. This permits the seats to be made of a variety of materials and be replaced if worn.

Globe Valves (13)

Globe valves serve primarily as throttling devices but can also be used for off-on service. The globe valve exhibits a high pressure drop as compared to gate valves due to the S-shaped path through the valve and valve disc being continually in the flow path.

Body, Bonnet and Stem — The comments applied to gate valve body, bonnets and stems are equally applicable to globe valves.

Discs — The disc in a globe valve is in a horizontal position as opposed to its vertical position in the gate valve. The globe valve disc can be flat, spherical or plug shaped. (Figure 6.11).

The flat disc is usually fitted with a flat insert of a compressible material. This combination is utilized to ensure tight shut-off while still maintaining a throttling characteristic. The insert material can vary from a simple composition material to PTFE. The choice of the insert material is dependent on the fluid conveyed and the temperature of the fluid.

The spherical and plug type discs both provide specific throttling characteristics. A metal to metal seal between the spherical or plug type disc and the valve seat enables the valve to be closed tightly. '

Butterfly Valves (9,14,15)

Butterfly valves are quarter turn valves that are used both for blocking and throttling service. (Figure 6.12). This is accomplished with the use of a movable metal disc in the valve which can be oriented in any position from parallel to the fluid flow wide open to 90° to the flow (closed). The valve pressure ratings can be as low as 16 psig to as high as 720 psig at ambient temperatures.

Size and Pressure Rating — Butterfly valves sizes can range from as

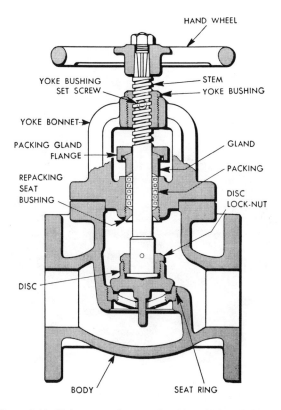

Figure 6.11. Globe valve. Courtest Stockham Valves & Fittings.

small as ¼ ″ to as large as 96″. The valves are generally rated in two ways. The first is the body rating which depends on the material of construction and temperature. The second is the differential pressure across the disc in the closed position. It is not unusual for both of these ratings to be the same. Ratings of 150 or 175 psig for a Class 125 or Class 150 body rating are common.

Bodies — Valve bodies can be made from bronze, cast iron, ductile iron, cast steel, stainless steel, aluminum or plastic material. The body can be wafer type, lug type, flanged, threaded or a number of other end connections such as grooved for use with a bolted coupling.

Wafer and lug style (also called single flange) bodies have gained wide acceptance. The wafer style body is bolted between pairs of flanges with flange bolts being outside the valve body. The lug body valve is bolted to the mating flanges. One problem with the wafer style is that if the piping on one side of the valve is removed, the valve will not stay in the line as the bolts holding it in

Figure 6.12. Butterfly valve. Courtesy Demco, Inc.

place must be removed. The lug style valve being contained by bolts from each of the mating flanges to threaded lugs will remain in position even if the piping on one side of the valve is removed. It is good practice to temporarily place a blind flange on the end of the valve when the pipe is to be removed.

An argument for the use of butterfly valves is that when compared to other valves that are available in comparable sizes, especially the larger sizes, the butterfly valve weighs less, costs less and does not require as extensive a support system. This is particularly true for the wafer and lug style bodies. The use of the flanged bodies detracts somewhat from the advantage but does not override it.

Threaded body butterfly valves are made in sizes ¼″ through 2″. These valves are not repairable.

Seats — Valve seats can be divided into two general categories, elastomeric or polymeric and metallic.

Elastomeric or Polymeric Seats — Elastomers such as Buna-N, EPDM or Neoprene are used as seat materials. The material is applied to the inside diameter of the valve bodies in one of three ways. The first is through the use of a removable flexible liner or boot. The second uses a removable reinforced insert or cartridge. In the third method the liner is fixed to the valve body (not removable).

Polymeric seats consist of a Teflon lining over an elastomer backup strip to provide sufficient resilience for sealing purposes. Teflon lined seats are made in both flexible boot and cartridge design seats.

Another form of plastomeric or polymeric seat involves the use of a narrow ring in the valve body against which the disc seats. The ring is mechanically held in place. Teflon is the usual material of construction for the ring in the "High Performance" type of butterfly valves.

Disc and Stem — The valve disc can be the same material as the valve, a different material, or lined with an elastomer or polymer. Two variations of the plain metal disc are the discs with an elastomer "O" ring or a metal ring in the disc edge. Both are used with metal seats. The all metal construction is applicable for high temperature service. The combinations of seats and discs are shown in the Table 6.19.

Table 6.19. Seats and Discs of Butterfly Valves

Seat	Disc	Remarks
Metal	Metal	Throttling service only
Metal	Metal with metal edge ring	Minimum leakage, high temperature service
Metal	Metal with elastomeric edge ring	
Metal with Polymer Seat Ring	Metal	
Elastomer lined	Plastic	
Elastomer lined	Elastomer lined	
Polymer lined	Metal	
Polymer lined	Polymer lined	

Valve stems are generally made of a stainless steel with more corrosion resistant materials such as Monel or Hastelloy sometimes used. Whether the more corrosion resistant stem is worth the cost is subject to question. If the argument that the stem seals are sufficient to contain the fluid being handled is accepted, then the more corrosion resistant stem material is not required.

The valve stem is fixed to the disc in one of two ways. In the first one piece stem passes through the disc being fastened to the disc with bolts or pins. In the other, the stem is in two pieces engaging the top and bottom of the disc through square or hexagonal recesses or protruding rectangular lugs.

Seals — Butterfly valves are provided with both primary and secondary stem seals. The primary seals are generally at the seat outside diameter stem and body inside diameter interface or at the seat inside diameter stem and interface. The design of this seal is as varied as are manufacturers of butterfly valves. The seal could be affected by the use of an "O" ring, square cross section ring or a raised section on the disc (a hub) in contact with the elastomer seat.

The secondary seals can be conventional packing, "O" rings or a cylinder of packing material. The packing may or may not be adjustable. If the secondary seal is found to be leaking it indicates that the fluid is getting through the primary seal area. Should the fluid being handled be corrosive and the valve seat and disc but not the body be resistant to the fluid, the stem housing and possibly the stem will be subject to the corrosive action of the fluid.

Operators — There are two basic types of actuators, the lever type and the gear operator. As the valve size and the volume of fluid handled increase, the force needed to operate the valve also increases. As a general rule, gear operators should be used on valves 6″ and larger.

Should the lever handle not provide degree of modulation required or the closure time of valve must be controlled to avoid the effects of water hammer, then the gear operators should be used regardless of valve size.

Gear operators are available in either open or enclosed types. The enclosed gear type has the advantage for the CPI that the gears are located in a housing which reduces or eliminates environmental effects. Gear operators are not usually provided with position locking devices as the gear system inhibits free rotation of the disc. Any type of device suitable to operate the valve can be attached to the stub shaft off the gear operator. A round hand wheel is used quite often. A chainwheel can be used where the valve is above an accessible position. An extension shaft can be used where a handwheel is preferred over a chainwheel. At least one intermediate support to maintain alignment of the extension shaft and the gear operator stub shaft should be considered.

Ball Valves (16,17)

Ball valves are compact quarter turn valves (open to close) used in blocking and throttling service. (Figure 6.13). They are relatively simple in construction consisting of the body, ball, stem, seats and stem seal.

Body Construction — Valve bodies can be constructed of a wide variety of materials. Among those are bronze, ductile iron, cast steel, stainless steel, nickel, titanium, PVC, CPVC, polypropylene, PVDF and fiberglass reinforced plastics. One valve design incorporates a lining of PTFE on the ball and inside of the valve body.

Threaded valve bodies are rated from pressures of 150 psig for the plastic to 600 psig or higher for the metallics. These pressure ratings are at ambient temperatures and decrease with increasing temperature.

Flanged valve bodies usually carry a more restricted rating, 150 psig and 300 psig for metallic valves and 150 psig for plastic valves. The flanges on metallic, FRP and one manufacturer's plastic valves are integral with the valve bodies. But for most of the flanged plastic valves available, the flanges are attached to the valves with short nipples through solvent or thermal welding.

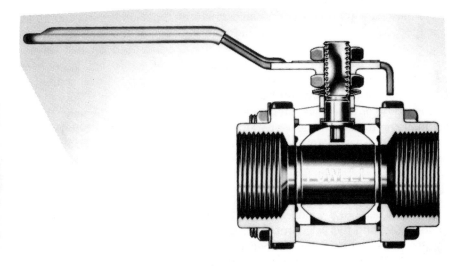

Figure 6.13. Ball valve. Courtesy Wm. Powell Co.

Flanged and threaded end metallic valves have several similarities in design. There are one, two and three piece valve bodies. From the standpoint of assembly of valve internals, the bodies are classified as end or top entry.

The end entry, metallic and plastic, valves are constructed for access to the primary part of the valve body through one or both ends of the valve. This is accomplished with the use of a threaded insert, one or two union ends, 2-piece or 3-piece bolted bodies. The top entry valve has a bolted top plate for access to the valve internals.

Ball and Stem — The valve ball material can be the same material as the valve body. In some instances the ball is a different material for corrosion prevention or chrome plates for erosion reduction. The stem material is generally the same as the ball.

The ball is held in place by the seats and the stem which rides in a slot in the top of the valve. There are several designs in which the ball and stem are one piece instead of two pieces.

The stem and stem housing should provide the valve with a blowout-proof stem feature. This avoids the possibility of the stem being forced out of the valve under pressure.

Seats and Seals — The seats and seals are usually PTFE with glass reinforced PTFE the normal seat material by some manufacturers and an alternate by others.

The stem seals consists of two parts, a lower seal and the upper seal. The lower seal is a washer shaped piece of PTFE material. The upper seal can be the same or it can be conventionally shaped PTFE packing of one or more pieces. Which packing system is best is argumentative. Pressure is maintained on the packing through a packing adjustment arrangement which can be as simple as either a packing nut threaded into the stem or the valve operator.

Operators — Lever and gear type operators are used on ball valves. Through the 4″ or 6″ size lever operators are suitable. For 8″ and larger gear operators should be used. For smaller valves, say through 1″, oval shaped or the lever handle in the shape of a "C" or "T" are also suitable. As the ball valve is a quarter turn valve, stops are usually incorporated in the valve top to limit the lever travel. The gear operator would have travel stops incorporated into its design. Gear operators can be provided with plain wheel or chain wheel handles.

Plug Valves (18)

The basic parts of the plug valve are the body, plug, the top works carrying the stem seal and the valve operator. (Figure 6.14 and 6.15). A quarter turn operates the valve from the opened to the closed position. While the valve may not be intended for throttling it is used for this purpose quite often. Another feature of this valve is its compact design as compared to gate or globe valves.

Material of Construction and Pressure Rating — Plug valves of metallic construction are used in the CPI. They are rated as low as 125 psig for screwed valves to 720 psig (Class 300) and higher for flanged valves. Brass, cast iron, ductile iron, cast steel, stainless steel, nickel, monel and titanium are among the metals from which the valves can be manufactured.

In addition to the all metal valves, plastic lined metal valves find extensive use. The plastics used as lining materials include Saran, polypropylene, PVDF, FEP, PTFE and PFA. The pressure rating of lined valves is the same as that of the metal bodies but the temperature rating depends on the plastic lining, from 175°F for Saran to 400°F for PFA. The corrosive nature of the fluid can reduce the useful temperature of the linings below 400°F except for the FEP, PTFE and PFA materials.

Lubricated Versus Non-Lubricated — There are two basic plug valves designs. One is the lubricated and the other non-lubricated design. A considerable amount of force would be required to turn the plug in the valve body unless the friction between the mating surfaces between the plug and body is reduced.

Figure 6.14. Plug valve. Courtesy Xomox, Tuftline Div.

In the lubricated valve design, a lubricant is forced through the valve stem to the plug-body interface. The lubricant reduces friction and permits easier valve operation. A high temperature lubricant permits valve operation at elevated temperatures. The disadvantage with this design is that the valve must be lubricated to properly operate.

The non-lubricated design utilizes a plastic interface between the plug and body to reduce operating friction. There are two basic versions of the non-lubricated plug valve. One utilizes a sleeve of PTFE material between the plug and body. The other design involves an all plastic lined design where both the body interior and plug are lined. The first design provides a lesser cost except if materials other than ductile iron and carbon steel are required for corrosion reasons.

The choice of all plastic lined or corrosion resistant metallics with PTFE sleeve design is mostly a matter of economics and availability.

There are at least two other variations in the non-lubricated valve design. In the first, two PTFE cylinders contoured to the plug face are utilized instead of the sleeve. The second involves the use of a plug design which could be described as "half a plug." For tight shutoff the plug seating surface is lined with an elastomeric material. The body can also be lined for erosive or corrosive service. For high temperature service the plug and body are not lined but tight shutoff may not be obtained.

Body — The valve body with flanged ends is available in multiport configurations. While the two port design (straight through flow) predominates, 3, 4 and 5 ported valves are available for flow diversion purposes. Screwed

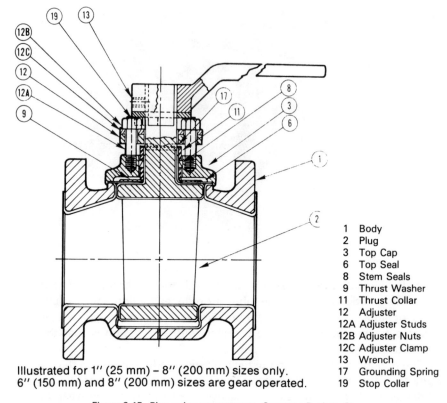

Illustrated for 1'' (25 mm) – 8'' (200 mm) sizes only.
6'' (150 mm) and 8'' (200 mm) sizes are gear operated.

1	Body
2	Plug
3	Top Cap
6	Top Seal
8	Stem Seals
9	Thrust Washer
11	Thrust Collar
12	Adjuster
12A	Adjuster Studs
12B	Adjuster Nuts
12C	Adjuster Clamp
13	Wrench
17	Grounding Spring
19	Stop Collar

Figure 6.15. Plug valve components. Courtesy Duriron Co.

end bodies are usually limited through the 2'' size.

Stem Seals — The stem seals vary from the unique design to the conventional. PTFE is a common packing material either by itself or in conjunction with other materials.

Operators — A rule of thumb for operator selection is to use the lever operator through the 3'' size valve and the gear operator for larger valves. The torque required to operate the valve limits the usefulness of the lever as they become quite long for the 3'' valve, up to 2'. Gear operators can be similar to those described under butterfly valves.

Diaphragm Valves

Diaphragm valves differ from other valves in that the body is isolated from the bonnet section by a diaphragm. (Figure 6.16). The diaphragm also serves as the closure element being extended into the valve body. The use of a flexible

Figure 6.16. Diaphragm valve. Courtesy Hills McCanna Co.

diaphragm does limit the operating pressure at which the valve can be used. As the diaphgram materials do not have an indefinite life, when the diaphragm fails stem leaks will occur and the valve will not function properly.

Rating — The smaller valves (½ ″ — 1 ″) can be used at pressures as high as 200 psig at or near ambient temperatures. For larger valves (14″ to 20″) the maximum pressure falls to 50 psig. As the temperature increases, the pressure characteristics decreases for all size valves.

Body — Two body configurations are available. One exhibits an inverted "U" flow pattern and the other a straight through pattern. The straight through pattern has advantages for handling viscous fluids and slurries or where an unobstructed flow path and reduced pressure drop are required. But for economic reasons the use of the inverted "U" pattern valve predominates.

Unlined valve bodies are made of most castable metals or moldable plastics. The less corrosion resistant metal bodies can be lined with elastomers (natural rubber, neoprene, etc.), polymers (Saran, Kynar, etc.) or glass. Linings are available only on flanged valve bodies.

Metal valve bodies are made with screwed, socket, buttweld and grooved end connections. Screwed, socket and flanged end connections are available on plastic bodies.

Diaphragms — Diaphragms are made from natural rubber and a wide variety of synthetic materials such as Buna N, ethylene, propylene and Viton. PTFE lined diaphragms are made for the more corrosive applications. The diaphragm material temperature limitations set maximum operating temperature for the valve.

Bonnet, Stem and Operator — The valve bonnet need not be the same material as the valve body as it does not come in contact with the material being handled. The bonnet is usually cast iron. One manufacturer of plastic body valves uses the same material for the bonnet.

The stems are basically the rising stem-inside screw type. To the bottom of the stem is attached the (diaphragm) compressor which in turn is attached to the diaphragm. The compressor is usually circular in shape and convex toward the diaphragm. Fingers extend out from the compressor center to evenly distribute the force applied to the diaphragm to obtain tight closure.

The valves can be equipped with an indicating or non-indicating stem. The indicating stem provides the operator with an indication of the diaphragm position. The indicating stem can be equipped with travel stops to limit the range of the diaphragm travel.

Handwheels predominate as manual operators. Auxiliary devices such as chainwheels and stem extensions are also available.

Pinch Valves

Pinch valves find use in handling fluids that require an uninterrupted flow path (e.g., slurries) and/or are corrosive. Leak paths are minimized by the use of a flexible conduit through which the material flows.

This valve operates through the collapsing of a flexible conduit. The flexible conduit is one of the two parts of the pinch valve. Through the other part, force is applied to collapse the conduit. (Figure 6.17).

The flexible tube (or double diaphragm) can be made of natural rubber, synthetic elstomers or PTFE. The valve can be operated through the application of hydraulic, pneumatic or mechanical pressure. In the case of the use of hydraulic or pneumatic pressure, a metallic enclosure is necessary. For mechanical actuation an enclosure is not necessary but may be required to contain the fluid being handled in case of leakage through the tube.

The pressure and temperature rating of this type of valve are limited. Construction limits the temperature to about 220 °F and the pressure to about 100 psig. Valve sizes run from 1″ to about 14″. All these valves are flanged.

Check Valves (19,20,21)

A wide variety of check valves are available as is shown in the Table 6.20.

Pressure ratings to 10,000 psig are possible. But most of the check valves are in the lower pressure categories Class 125, Class 150, etc. The valve's

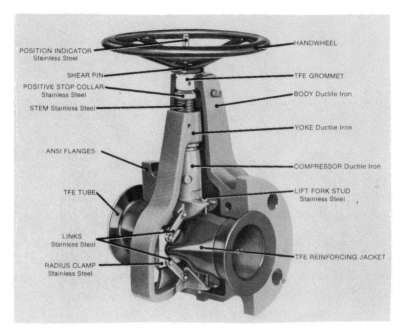

Figure 6.17. Pinch valve. Courtesy Resistoflex Corporation.

Table 6.20. Design of Check Valves

Body	Single Disc	Double Disc	Check Ball	Piston	Flexible Tube
Screwed	X		X	X	X
Flanged	X	X	X	X	X
Wafer/Lug	X	X			
Insert				X	

bodies are available in a wide variety of materials, e.g., bronze, cast iron, steel, stainless steel, nonferrous materials, plastic lined steel and thermoplastic materials. The valve internals are available in almost as many materials plus thermosetting plastic. The valve sizes cover the range ¼ " through 24" with screwed valves covering the lower end and the flanged and wafer construction covering a very broad range, (1"–24"). A typical screwed swing check valve is shown in Figure 6.18. Lift type check valves are illustrated in Figures 6.19 and 6.20.

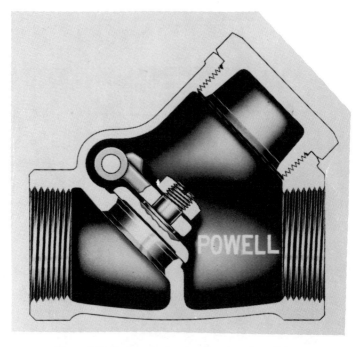

Figure 6.18. Swing check valve. Courtesy Powell Co.

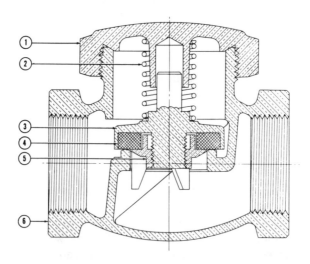

(1) Cap
(2) Spring
(3) Disc holde
(4) Disc
(5) Disc guide
(6) Body

Figure 6.19. Spring loaded lift check valve. Courtesy Jenkins Valves.

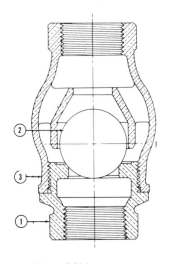

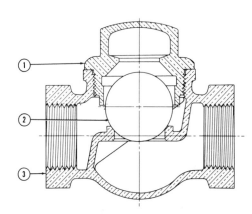

Figure 6.20A   Figure 6.20B

(1)Cap hub (2) Ball (3) Body

Figure 6.20A and 6.20B. Ball lift check valves for (A) vertical piping and (B) horizontal piping. Courtesy Jenkins Valves.

SUPPORTS, GUIDES AND ANCHORS

Suggested support spacing for uncomplicated metallic piping systems at moderate temperatures are shown in Table 6.21. Support spacing for plastic piping (22) are shown in Table 6.22. Types of supports and hanger are shown in Figure 6.21 (23). Point supporting of plastic piping should be avoided as it causes a concentration of weight and could lead to premature pipe failure. To distribute the weight at a support point the use of a sleeve of medium gage sheet metal or plastic covering the bottom 1/3 to 1/2 of the pipe is suggested.

The following standards deal with support systems.

MSS SP-58 Pipe Hangars and Supports — Materials, Design and Manufacture.

MSS SP-69 Pipe Hangars and Supports — Selection and Application.

Anchors and guides are used to control pipe movement. This movement may be due to thermal effects, wind, snow, or ice, seismic activity, or safety device operation. Placement of guides and anchors can be determined by previous experience. If previous experience cannot be utilized, a system of calculations can be used to determine if there is sufficient flexibility in the system to avoid overstressing the piping or the equipment to which it is attached.

Table 6.21. *Suggested Support Spacing for Metallic Pipe*

(1,500-psi stress, 1/10-in. deflection, water-filled pipe)

Nominal pipe size	1	1½	2	2½	3	3½	4	5	6	8	10	12	14	16	18	20	24
Span	7	9	10	11	12	13	14	16	17	19	22	23	25	27	28	30	32

Table 6.22. *Support Spacing for Plastic Piping*

NOMINAL PIPE Diam, inches	POLYVINYL CHLORIDE (PVC)								CPVC				POLYPROPYLENE				FIBERGLASS REINFORCED				KYNAR®*			
	SCH 40				SCH 80				SCH 80				SCH 80				200 PSIG				SCH 80			
	60 °F	100 °F	140 °F	180 °F	60 °F	100 °F	140 °F	180 °F	60 °F	100 °F	140 °F	180 °F	60 °F	100 °F	140 °F	180 °F	60 °F	100 °F	140 °F	180 °F	80 °F	100 °F	140 °F	160 °F
	Support Spacing feet																							
½	4½	4	2½	DO NOT USE FOR THIS TEMPERATURE	5	4½	2½	DO NOT USE FOR THIS TEMPERATURE	5½	5	4½	2½	4	3		CONTINUOUS SUPPORT	Not Available				4½	4½	2½	CONTINUOUS SUPPORT
¾	5	4	2½		5½	4½	2½		6	5½	4½	2½	4	3			Not Available				4½	4½	3	
1	5½	4½	2½		6	5	3		6½	6	5	3	4½	3			9	8½	8½	7½	5	4¾	3	
1½	6	5	3		6½	5½	3½		7	6½	5½	3½	5	3½	2		9	8½	8½	7½	5½	5	3	
2	6	5	3		7	6	3½		7½	7	6	3½	5	3½	2		9	8½	8½	7½	5½	5¼	3	
3	7	6	3½		8	7	4		9	8	7	4	6	4	2½		10	9½	9½	9	Not Available			
4	7½	6½	4		9	7½	4½		10	9	7½	4½	6	4½	3		11	10½	10½	9	Not Available			
6	8½	7½	4½		10	9	5		11	10	9	5	6½	5	3		12	12	11½	11	Not Available			
8	9	8	4½		11	9½	5½		Not Available				7	5½	3½		13½	13½	12½	11½	Not Available			
10	10	8½	5		12	10	6		Not Available				Not Available				16	15½	13½	13	Not Available			

Figure 6.21. Types of pipe hangers. Courtesy Manufacturers Standardization Society.

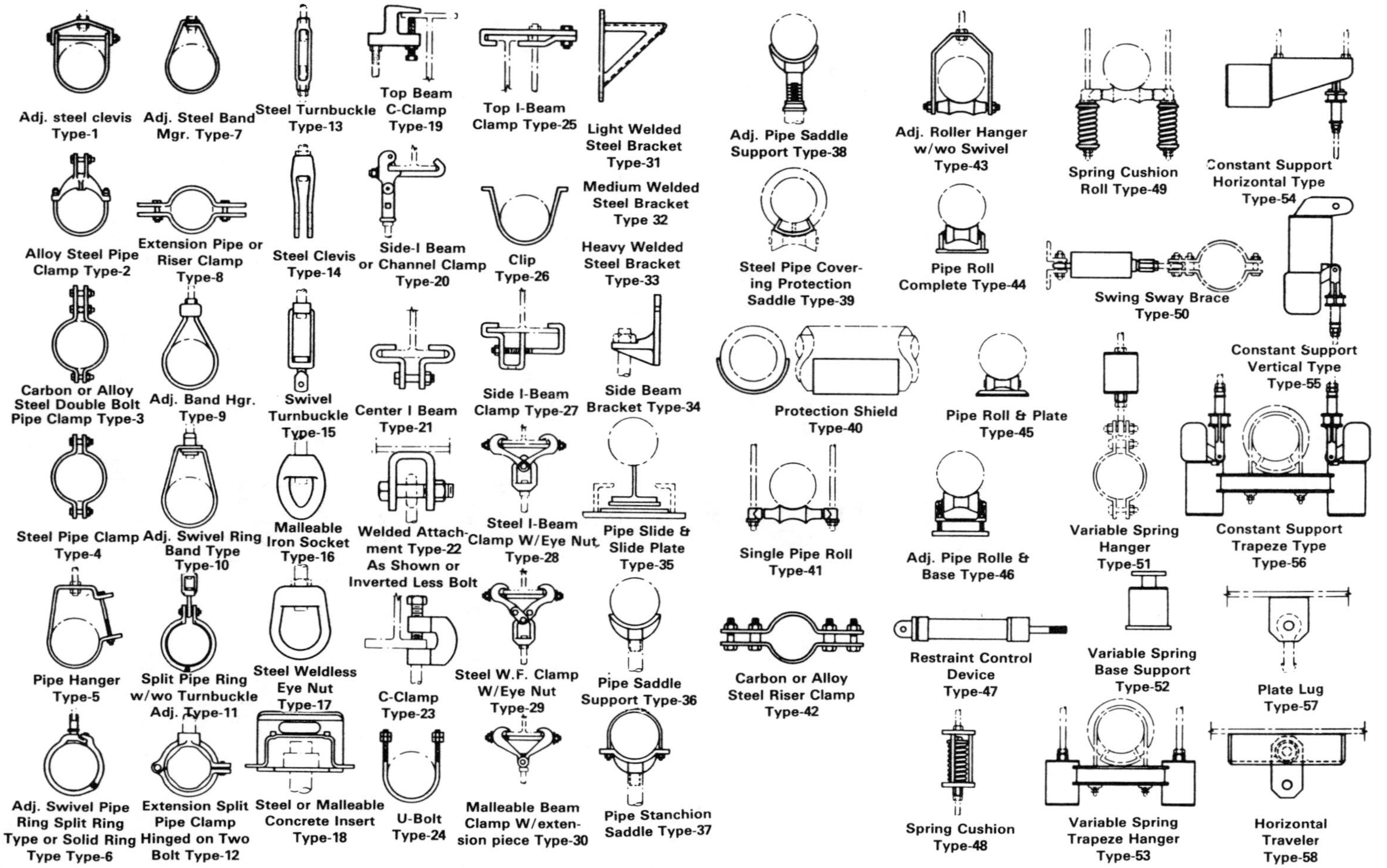

Constant Support Horizontal Type-54

Constant Support Vertical Type-55

Constant Support Trapeze Type-56

Plate Lug Type-57

Horizontal Traveler Type-58

Spring Cushion Roll Type-49

Swing Sway Brace Type-50

Variable Spring Hanger Type-51

Variable Spring Base Support Type-52

Variable Spring Trapeze Hanger Type-53

Adj. Roller Hanger w/wo Swivel Type-43

Pipe Roll Complete Type-44

Pipe Roll & Plate Type-45

Adj. Pipe Rolle & Base Type-46

Restraint Control Device Type-47

Spring Cushion Type-48

Adj. Pipe Saddle Support Type-38

Steel Pipe Covering Protection Saddle Type-39

Protection Shield Type-40

Single Pipe Roll Type-41

Carbon or Alloy Steel Riser Clamp Type-42

Light Welded Steel Bracket Type-31

Medium Welded Steel Bracket Type 32

Heavy Welded Steel Bracket Type-33

Side Beam Bracket Type-34

Pipe Slide & Slide Plate Type-35

Pipe Saddle Support Type-36

Pipe Stanchion Saddle Type-37

Top I-Beam Clamp Type-25

Clip Type-26

Side-I Beam Clamp Type-27

Steel I-Beam Clamp W/ Eye Nut Type-28

Steel W.F. Clamp W/Eye Nut Type-29

Malleable Beam Clamp W/extension piece Type-30

Top Beam C-Clamp Type-19

Side-I Beam or Channel Clamp Type-20

Center I Beam Type-21

Welded Attachment Type-22 As Shown or Inverted Less Bolt

C-Clamp Type-23

U-Bolt Type-24

Steel Turnbuckle Type-13

Steel Clevis Type-14

Swivel Turnbuckle Type-15

Malleable Iron Socket Type-16

Steel Weldless Eye Nut Type-17

Steel or Malleable Concrete Insert Type-18

Top Beam C-Clamp Type-19

Extension Pipe or Riser Clamp Type-8

Adj. Band Hgr. Type-9

Adj. Swivel Ring Band Type Type-10

Split Pipe Ring w/wo Turnbuckle Adj. Type-11

Extension Split Pipe Clamp Hinged on Two Bolt Type-12

Adj. steel clevis Type-1

Alloy Steel Pipe Clamp Type-2

Carbon or Alloy Steel Double Bolt Pipe Clamp Type-3

Steel Pipe Clamp Type-4

Pipe Hanger Type-5

Adj. Swivel Pipe Ring Split Ring Type or Solid Ring Type Type-6

Adj. Steel Band Mgr. Type-7

The subject of piping flexibility analysis is covered extensively in the literature (24,25). As the calculations can become very involved, computer programs have been written to overcome the time required in manual calculation.

PRESSURE TESTING

Piping systems are tested to prove their integrity. Both liquids and gases are used as the test medium with the use of liquid preferred. The test pressure should be based on the pressure for which the piping system is designed to withstand. The reader is referred to ANSI B31.3 para. 337 for a full discussion of pressure testing.

Hydrostatic testing is carried out at one and one-half times the system design pressure. Water is the usual test medium. The test should be maintained for a sufficient period of time to determine the absence of leaks. Should leaks occur they are to be repairs and the system retested.

Pneumatic testing is carried out at 110% of the piping system design pressure. Air is the usual test medium. Extra precautions should be taken when pneumatic testing due to the possible release of compressed air accompanied by high velocity and an expansion in volume. The test pressure should be increased gradually in steps so should a leak exist it can be found at a lower rather than higher pressure. Pneumatic testing should be used only when it be prohibitive to admit water to the piping system.

Typical Specifications

The following is a check list of what items can be included in the typical specification.

1. *Pipe* — material and dimension spec (ASTM A53 Type S Grade B, ASTM A106 Grade B etc.) and schedule (thickness) Sch. 40, Sch. 80, etc.

2. *Threaded fittings* — material cast iron, malleable iron, etc., to material spec (ASTM A126 Grade B, ASTM A97, etc.), dimension spec, (B16.4, B16.3 etc.) thread spec (B2.1). For threaded unions specify seat materials if different than body material.

4. *Butt weld fittings* — dimension spec (ANSI B16.9) material spec (ASTM A216, Grade WPB, etc.) and schedule.

5. *Flanges* — dimension spec (ANSI B16.1, B16.5, etc.), pressure spec. (Class 150, Class 300, etc.), material spec (ASTM A105, etc.), flange type (weld neck, slip on, etc.), flange face (raised face, tongue groove, etc.). If the welding neck flange is specified, the wall thickness of the pipe to which it is attached must also be specified.

6. *Flanged fittings* — dimension, pressure and material specs similar to those for flanges.

7. *Bolts and nuts* — material specs (ASTM A307 Grade B, ASTM A193

Grade 17, etc.), specify machine or stud bolts, regular hex or heavy hex pattern bold head and nuts.

8. *Gaskets* — type (compressed asbestos, spiral wound, etc.), thickness, form (ring type, full face, etc.), ASTM or brand designation.

9. *Welding* — ANSI B31.3, ASME B(PV Section IX or AWS — site appropriate sections, including welding inspection.

10. *Branch connections* — requirements such as size limitations, construction and reinforcement requirements.

11. *Valves* — include valve descriptions or valve ID number (which refers to another list for the valve description).

The service or services to which the specification call apply can be included in the specification with tne pressure and temperature requirements.

REFERENCES

1. Ludwig, E. G., "Applied Process Design For Chemical and Petrochemical Plants" Vol. 1, p. 54.
2. National Bureau of Standards, Product Stanard NBS PS)-15-69, pg. 8.
3. Stockham Valves and Fittings Catalog.
4. Issac, M. and Sutterlund, R. B., "A Guide to Joining Methods" Chemical Engineering, May 3, 1971.
5. Kent, G. R., "Selecting Gaskets for Flanged Joints," Chemical Engineering, March 27, 1978.
6. Cunningham, E. R., "Mechanical Pipe Couplers," Plant Engineering, Sept. 14, 1978.
7. Pikulik, A., "Selecting and Specifying Valves For New Plants," Chemical Engineering, Sept. 13, 1976.
8. Wier, J. T., "Selecting Valves For CPI," Chemical Engineering, Nov. 24, 1975.
9. Evans, F. L., "Valves For HIP," Hydrocarbon Processing, June 1974.
10. Schweitzer, P. A., "Handbook of Valves," Industrial Press, 1972.
11. Lyons, J. L., and Ashland Jr., C. L., "Lyons Encyclopedia of Valves," Von Nostrand Rheinhold Co., 1975.
12. Edfeldt, P., "Valve Applications: IV, Gate Valves," Heating/Piping/Air Conditioning, Sept. 1977.
13. Pikulik, A., "Valve Applications: I, Globe Valves," Heating/Piping/Air Conditiong, March 1977.
14. Pannkoke, T. and Ryan, C. J., "Valve Applications: VI, Butterfly Valves," Heating/Piping/Air Conditioning, Jan. 1978.
15. Bertrem, B. E., "Butterfly Valves for Flow of Process Fluids," Chemical Engineering, Dec. 20, 1976.
16. O'Keefe, W., "H-p, h-t Ball and Butterfly Valves," Power, May 1977.
17. Pannkoke, T., and Ryan, C. J., "Valve Applications: V, Ball Valves," Heating/Piping/Air Conditioning, Nov. 1977.
18. "Valve Applications: II, Plug Valves,." Heating/Piping/Air Conditioning, May 1977.

19. O'Keefe, W., "Check Valves," Power, August 1976.
20. Micholson, G. D., "Why the Swing to Wafer Type Check Valves," Hydrocarbon Processing, June 1974.
21. Pannkoke, T., and Ryan, C. J., "Valve Applications: III, Check Valves," Heating/Piping/Air Conditiong, July 1977.
22. Chasis, D. A., "Plastics Piping Systems," Industrial Press, 1976, pg. 105.
23. Manufacturers Standardization Society, Standard Practice SP69, pg. 4.

24. "Design of Piping Systems," The M. W. Kellogg Company.
25. "Piping Design & Engineering," ITT-Grinnell Company.

BIBLIOGRAPHY

Crocker, S., and King, R. C., "Piping Handbook," McGraw-Hill, 5th Ed., 1967.

Evans, Jr., F. L., "Equipment Design Handbook for Refineries and Chemical Plants," Vol. 2, Chaper 7, Gulf Publishing Co., 1964.

Holmes, E., "Handbook of Industrial Pipework Engineering," John Wiley & Sons, 1963.

Mallison, J. H., "Chemical Plant Design with Reinforced Plastics," McGraw-Hill, 1969.

Rase, H. F., "Piping Design for Process Plants," John Wiley and Sons, 1963.

"Piping Guide," Syentek Books, 1973.

Tube Turns, "Piping Engineering," 1975.

Weaver, R. B., "Process Piping Design," 2 vols., Gulf Publishing Co., 1973.

SELECTED REFERENCES FOR PIPE SIZING

"Cameron Hydraulic Data," Ingersoll-Rand Company.

"Flow of Fluids Through Valves, Fittings, and Pipes," Crane Company.

Kent, G. R., "Preliminary Pipeline Sizing," *Chemical Engineering,* Sept. 25, 1978.

Kern, R., "Useful Properties of Fluids for Piping Design," *Chemical Engineering,* Jan. 8, 1975.

Kern, R., "How to Design Piping for Pump-Suction Conditions," *Chemical Engineering,* Apr. 28, 1975.

Molte, C. B., "Optimum Pipe Size Selection," Trans Tech Pub., 1978.

Index